Ice Age Floodscapes of the Pacific Northwest

Bruce Norman Bjornstad

Ice Age Floodscapes of the Pacific Northwest

A Photographic Exploration

 Springer

Bruce Norman Bjornstad
Richland, WA, USA

ISBN 978-3-030-53045-7   ISBN 978-3-030-53043-3   (eBook)
https://doi.org/10.1007/978-3-030-53043-3

This Springer imprint is published by the registered company Springer Nature Switzerland AG
The registered company address is: Gewerbestrasse 11, 6330 Cham, Switzerland

Acknowledgments

This work benefitted immensely from thoughtful reviews by many highly respected and renowned geologists including Vic Baker, Richard Waitt, Jim O'Connor, Gene Kiver, Kevin Pogue, Daniel Mann, and Nick Zentner. Also, helpful in the field, were Andrea Balbas and Tom Cooney. Special thanks go to artist Stev Ominski who, over the last couple of decades, I have appreciated collaborating with to reproduce representative renditions of the Ice Age megafloods, many of which are illustrated herein. Lastly, thanks to Ron Doering at Springer Nature who spawned and supported the idea of producing a visual portrait of colossal floods that shaped the Pacific Northwest.

About the Book

This book is the culmination of 40 years of musings and scientific study upon the Ice Age floods by the author. Often, an appreciation of the huge scale of the Ice Age floods and the features they left behind are lost upon examination at close range. Thus, aerial surveys often provide the best perspectives for studying megafloods and features left behind. Thanks to recent advances in remote-controlled drone technology, geologists now have a useful new tool to study, understand, and appreciate the incredible scale and variety of landforms resulting from the Ice Age floods. The following pages hopefully will increase the reader's understanding and appreciation of the immense power and magnitude of these massive Earth-changing events that occurred so recently in the geologic past.

Contents

About the Author

Bruce Bjornstad is a licensed geologist/hydrogeologist and retired Senior Research Scientist from Battelle's Pacific Northwest National Laboratory living in Richland, Washington. He received a bachelor's degree in geology from the University of New Hampshire and a master's degree in geology from Eastern Washington University. During his 35-year career, he has written numerous documents and reports on the geology of the region as well as two geologic guidebooks titled "On the Trail of the Ice Age Floods" that transformed the Pacific Northwest as recently as 14,000 years ago. Bruce is also the creator of an online YouTube Channel titled: Ice Age Floodscapes.

Background

Pioneering Geologist – J Harlen Bretz

Clues to repeated devastation are written all over the landscape of the Pacific Northwest. Although individual flood features rarely provide conclusive evidence for large-scale floods, when examined collectively they tell an amazing tale of repeated cataclysms. It was J Harlen Bretz in 1923 who first proposed an "outrageous hypothesis" that a giant flood produced some very unusual landforms in the Channeled Scabland of eastern Washington. Over the next decade, he meticulously documented his observations and evidence for a "great flood of water" even though he couldn't explain where the water came from. Early on Bretz used the term "Spokane Flood" to describe the area from where the water appeared to come. During this time Bretz focused his attention on the field relations and geologic evidence for flooding itself, even though he was uncertain of the exact source of water.

Bretz spent much of his career attempting to convince a mostly skeptical geologic community of his Spokane Flood hypothesis. Others in the profession passionately attacked his ideas because they smacked too much of the Bible and were contrary to Uniformitarian principles. Uniformitarianism holds that geologic changes come about from slow, gradual, steady geologic processes – similar to rates we observe today. In short, Bretz's catastrophic flood was too abominable in other geologists' eyes at the time. Bretz was adamant in his defense for an Ice Age flood and was resolute that the evidence could not be convincingly explained any other way. Most of Bretz's critics, who never directly observed the evidence, went to great lengths to dismiss flood features as forming from normal glacial or other non-catastrophic processes.

J Harlen Bretz at age 67 (Julian Goldsmith photo)
"The only genetic interpretation yet proposed which is inherently harmonious and which fits all know facts is that of a great flood of water…" Bretz (1927)

Unfazed by his critics Bretz continued to examine and gather evidence for catastrophic outburst floods for four decades. Later, his ideas gained more acceptance based on confirming evidence viewed from the air like expansive fields of giant current ripples visible on aerial photographs within the Channeled Scabland. Elsewhere, evidence for the sudden, rapid draining of glacial Lake Missoula further cemented the Ice Age flood story. Finally, in 1979, only two years before his death, Bretz was awarded the Penrose Medal, the highest honor bestowed by the Geological Society of America. Finally vindicated, perhaps Bretz's proudest accomplishments were that he was finally viewed as a visionary and that at age 98 he had outlived all his critics.

B. N. Bjornstad, *Ice Age Floodscapes of the Pacific Northwest*, https://doi.org/10.1007/978-3-030-53043-3_1

Thanks to Bretz's challenge, geologists were forced to objectively reexamine their beliefs and accept that some of the most dramatic landscapes on earth result from short-lived, natural, catastrophic events. Bretz opened the door and inspired other geologists to consider the evidence for other catastrophes that produce dramatic change, such as meteor impacts and volcanic eruptions. Bretz and his theory of the Ice Age floods are testaments for the scientific method and highlight the importance of gathering facts, keeping an open mind, and constantly questioning the dogma of our times.

Since Bretz, many other respected geologists have contributed to our understanding of Ice Age megafloods. Those dedicated scientists and their published works are listed in the bibliography at the end of this book.

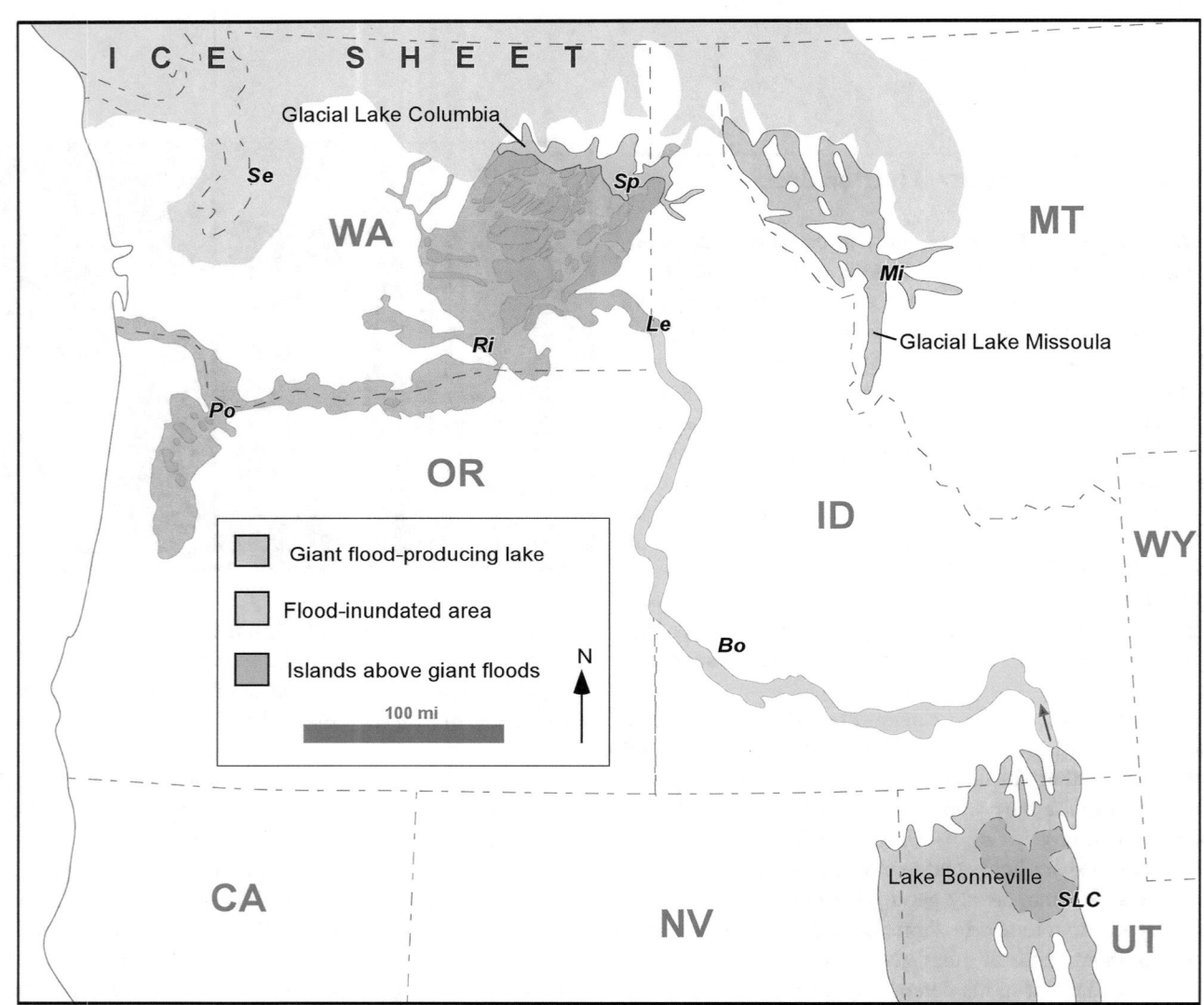

Three, Giant Ice-Age-flood Sources within the Pacific Northwest Giant Ice Age floods in the Pacific Northwest came from three different sources: glacial Lake Missoula, glacial Lake Columbia and pluvial Lake Bonneville. (Table 1) Se = Seattle, Sp = Spokane, Mi = Missoula, Le = Lewiston, Ri = Richland, Po = Portland, SLC = Salt Lake City.

Ice-Age-Flood Sources

Frequencies and relative magnitudes between the three flood sources are summarized in Table 1.

Glacial Lake Missoula – Up to 100, separate flood events during the last glacial cycle between 15 and 20 thousand years ago.

Glacial Lake Columbia – One flood occurred from the breakup of Glacial Lake Columbia, at the end of the last glacial cycle, about 14,000 years ago.

Pluvial Lake Bonneville – The Lake Bonneville flood entered the Columbia Basin from the east via the Snake River only once. Floodwater from Lake Bonneville downstream of Idaho was mostly confined to the lower Snake

Table 1 Ice-Age-Flood Sources in the Pacific Northwest. (Derived from O'Connor et al., 2020)

	Number of Flood Events During Last Glacial Cycle	Maximum Outflow Volume (mi³)	Maximum Outflow Volume (km³)	Peak Discharge Near Outlet (million ft³/sec)	Peak Discharge Near Outlet (million m³/sec)	Duration of Lake Drainage (days)	Maximum Outburst-Flood Duration (days)
Lake Missoula	~100	605	2540	1024–1235	15–30	~4	~20
Lake Columbia	1	118–145	55	≤35	~0.13	?	?
Lake Bonneville	1	1270	5130	39–57	0.9–1.6	~17	≥20

and Columbia river valleys – well away from the Channeled Scabland. Furthermore, because the flood lasted a couple weeks or more, instead of a few days like Lake Missoula, it was, by geologic accounts, less devastating even though twice as much water discharged from Lake Bonneville. Unlike the other sources there was no ice dam involved with this flood. It produced only one flood when an alluvial dam was overtopped at a low divide by rising level of Lake Booneville around the peak of the last Ice Age – around 18,000 years ago. During this time more water drained into the lake either because of more precipitation and/or less evaporation. The Lake Bonneville flood occurred between two of the many Lake Missoula megafloods.

A megaflood is considered a flood with a discharge greater than one million m³/sec (>35 million ft³/sec). The largest and most destructive floods clearly came from Lake Missoula – which discharged at >1200 million ft³/sec and, once initiated, drained in only a few days. Lake Bonneville flood lasted much longer (~17 days) but had a much lower discharge (<1.6 million m³/sec) compared to Lake Missoula and therefore was much less destructive. While Lake Columbia flood didn't officially qualify as a "megaflood" it still produced some significant flood features, like giant current ripples that are on par with the Lake Missoula floods.

Each of three flood sources will be covered separately in the order listed above for the remainder of this publication. Outburst floods for both Lake Columbia and Lake Bonneville, because of their smaller volume and discharge, were mostly confined to a single pathway. These will be examined separately, starting at their source, before covering their paths of destruction downstream.

Megaflood paths for outbursts from the much-larger glacial Lake Missoula, on the other hand, spread out for almost 100 miles across the Channeled Scabland. Flood features within four different scabland tracts (Cheney-Palouse, Telford-Crab Creek, Grand Coulee, and Moses Coulee), will be covered separately starting at the upper end and again moving downstream with respect to the floods.

This book focuses on the more obvious and most dramatic flood features where the evidence is most striking and convincing from the air. However, there are many other features and localities where the evidence for megafloods is more subtle and less visually impactful and therefore not included in this treatise. Again, the reader is referred to the list of previously published works listed in the back.

Foundations

Before diving into the subject of megafloods let's first establish the foundations upon which the floods imprinted themselves. Before the earliest Ice Age floods there were two older, widespread geologic units in eastern Washington, which huge floods encountered and significantly modified: (1) volcanic Columbia River basalt and (2) wind-deposited Palouse loess.

Columbia River Basalt

For more than 10 million years (6.5–17 million years ago), one flow after another of volcanic lava, known as the Columbia River basalt, inundated the region, creating the substrate for the broad, rolling plains of the Columbia Basin. Altogether this included as many as 300 lava flows, spreading out for hundreds of miles and causing the Earth's crust to sag under the massive weight of dense, basalt lava. Individual flows range up to 300 feet thick. The basalt flows cover about one-third of Washington state as well as large parts of Oregon, extending to Idaho and Nevada. Altogether, these basalts reached their maximum thickness (up to 15,000 feet) near the Tri-Cities area of south-central Washington but thin outward toward the margins.

Flood basalts in the Columbia Basin flowed or sometimes shot out from of many long, vertical and linear fissures mostly located in southeast Washington and northeast Oregon. Today, many of the fissures, or feeder dikes, that delivered lava to surface are visible across the

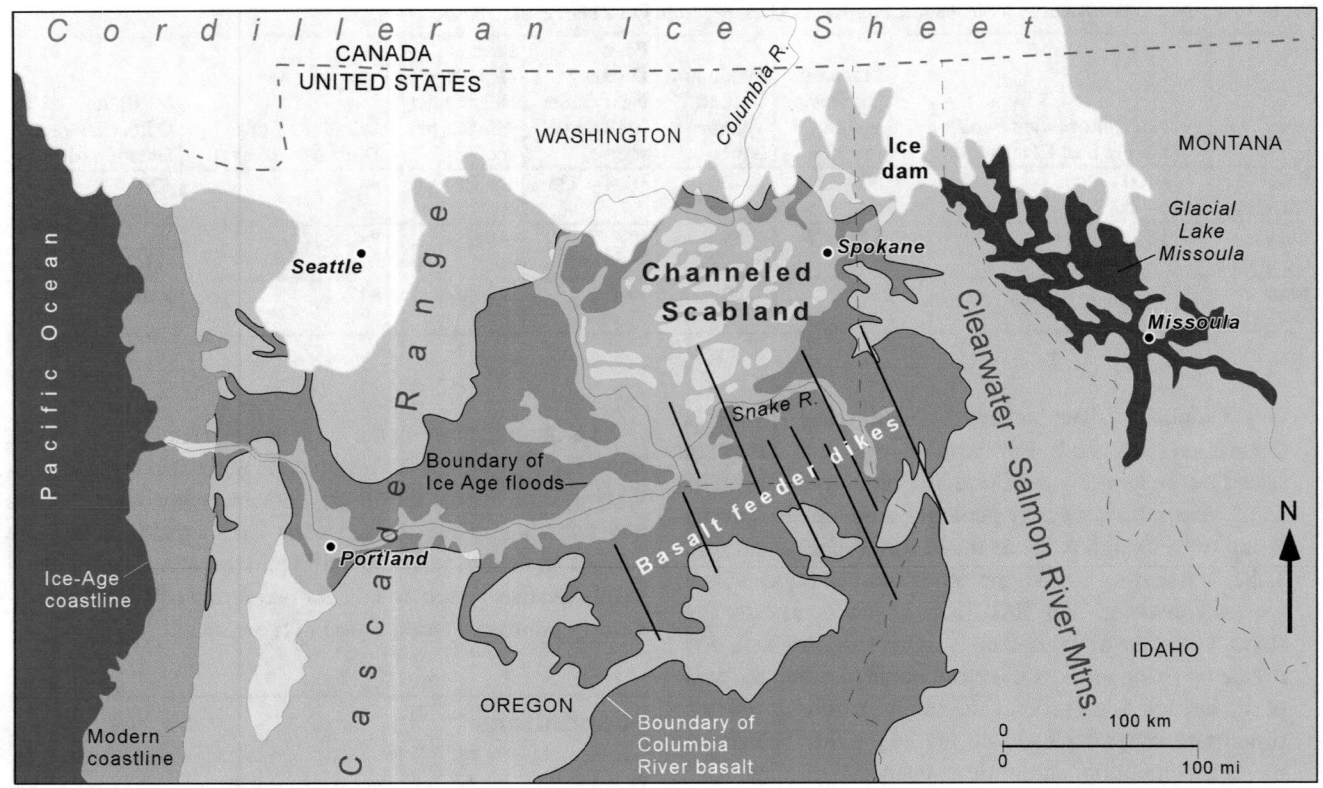

Floods of Basaltic Lava and Glacial Meltwater from Lake Missoula   Areas in Pacific Northwest were first affected by ancient floods of volcanic basalt (gray) and millions of years later by outbursts from glacial Lake Missoula (blue). Basalt lava came to the surface via parallel, deep underground fissures, also known as feeder dikes.

Columbia Basin. Then the molten lava flowed under the influence of gravity toward lowlands to the west and north. Based on distinctive chemical and physical characteristics, some of the same flows that originated in the higher, eastern part of the Columbia Basin can be traced hundreds of miles away to the Pacific Ocean. Basalt flows must have erupted rapidly and were extremely fluid based on: (1) the lack of any significant buildup by volcanic debris along the vents, and (2) distance traveled by many of the flows (i.e., hundreds of miles) before cooling and solidification. Apparently, an insulating, hardened crust developed at the top and bottom of each flow, which allowed the molten interior to continue flowing and spread outward for long distances.

The internal vagaries of basalt flows had a profound effect and played a significant role in how they were eroded by the Ice Age floods and the types of features and landforms left behind. A universal characteristic of basalt flows are fractures, or joints—shrinkage cracks that developed as the molten lava

cooled, contracted and solidified. Some of the thicker basalt flows may have taken as long as 50 years to cool and solidify. Towards the top of the flow, where cooling occurred more rapidly, a higher density of cracks developed in a random pattern called the "entablature" zone. Deeper within the same lava flow, where the lava was insulated beneath the hardening entablature zone, the lava remained molten for a longer period of time, and thus the basalt cooled more slowly. This allowed the growth of larger, more regular fractures that created a honeycomb-like network of polygonal columns referred to as the "colonnade". The cracks started at the base of the molten flow interior and slowly propagated upward towards the interior of the cooling lava flow. The end results are almost-continuous vertical columns spanning the entire breadth of the colonnade. Rubbly zones of broken-up rock may also be present both at the very tops and bottoms of flows. Oftentimes, spherical voids are concentrated near the tops of individual lava flows. These represent gas bubbles, frozen in place, during cooling of the molten lava flow.

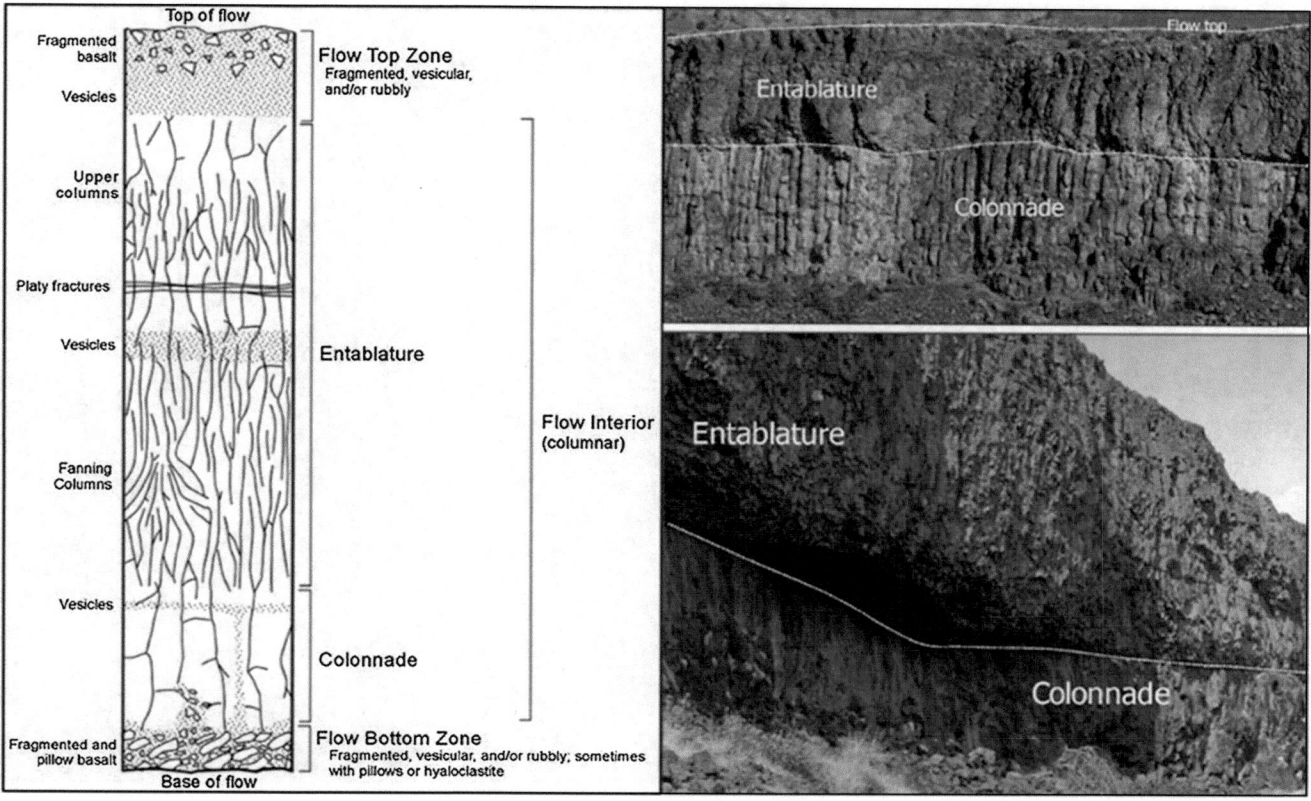

Internal Structure of a Basalt Flow   Considerable variation exists within and between individual basalt flows, but most of them are composed of at least two subunits: entablature and colonnade. Powerful Ice Age floods often preferentially eroded away larger blocks of basalt within the colonnade. The entablature proved to be more resistant to flood erosion producing rock overhangs like that above.

Stacking of Sheet-Like Lava Flows at Wallula Gap   Basalt flows at the base of this exposure are ~16 million years old – the youngest is 8.5 million years old. Many more flows lie buried below the level of the Columbia River.

Feeder Dike   Vertical fissure filled with Columbia River basalt cuts across several older lava flows along the Snake River. Dike was later exposed by downcutting of the river followed by erosion via Missoula floods.

Palouse Loess

Prior to the Ice Age floods, most of the Channeled Scabland was covered by a blanket of windblown sediment (loess) deposited onto rolling Palouse Hills across much of eastern Washington. The loess deposit began forming as predominantly southwest winds blew sediment up onto the Palouse Slope at the beginning of the Ice Age perhaps as far back as 2.6 million years ago. The earliest floods eroded this sediment and transported it downstream where it was redeposited in the slackwater Pasco, Othello and Quincy basins, when floodwaters backed up behind Wallula Gap. The dust storms were especially intense and effective at transporting sediment after each outburst flood, which removed and/or covered over the anchoring vegetation. Thus, the fresh blanket of flood sediment was easily picked up by the wind and carried back toward the Palouse region. The Palouse hills, then, are the end result of back-and-forth winnowing of flood sediment by desert winds over the last couple of million years.

Earliest Ice Age Floods

With the close of basalt volcanism, about 6.5 million years ago, only a few more million years elapsed before the region succumbed to a new era of flooding, this time of icy glacial meltwater. The Ice Age began with a period of distinct climatic cooling at the beginning of the Pleistocene Epoch, about 2.6 million years ago. The geologic record of flooding is best preserved for the last glacial cycle. Much less evidence exists for the earliest floods since the last floods tended to erode or cover up the evidence for older floods.

For at least the last million years or so, the Earth's climate has cycled between major periods of glacial advance and retreat, going through a complete cycle about every 100,000 years. While there is some evidence for floods going back a million years or more, most of that evidence was buried or destroyed by the last floods, which occurred during the last glacial cycle between 14 and 20 thousand years before present.

Rolling Palouse Hills Prior to Ice Age megafloods much of eastern Washington looked like this. These rolling Palouse hills, east of the Channeled Scabland, completely escaped erosion by the Missoula floods. Preserved is the gentle dendritic drainage pattern characteristic these mature hills of slowly accumulating windblown dust.

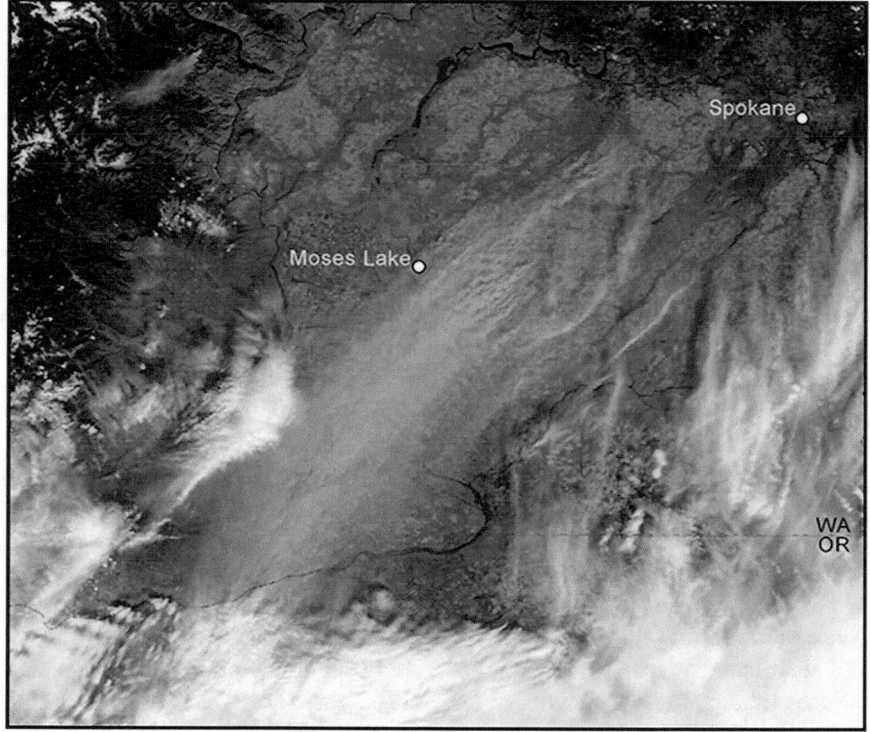

Dust in the Wind Satellite view of recent storm kicking up a brown cloud of dust (loess) that spreads out across the Channeled Scabland. Next page: Repeated periodic dust storms like these are responsible or the accumulation of up to 250 ft of loess over the last ~2 million years.

How do we know how old the floods are? One way geologists can identify the land record of old floods is by looking at the magnetic polarity of fine grained deposits that was imprinted onto the sediment as it was laid down. The Earth's magnetic field has periodically and randomly "flipped" back and forth over geologic time. The last significant reversal occurred about 780,000 years ago, when the magnetic field changed from a "reversed" direction to the "normal" direction we observe today. These changes often are preserved in sediments since some sediment grains are magnetic and, like tiny magnets, tend to align themselves with the Earth's magnetic field at the time they are deposited. So, early Pleistocene flood deposits retain a reversed magnetic polarity while middle Pleistocene and younger deposits record a normal magnetic polarity.

Evidence for the oldest floods on land can also be found by examining how flood deposits have weathered at the surface over time. Deposits from the last floods, which are relatively young (14,000–15,000 years), show little or no weathering. In contrast, old flood deposits exposed to the elements for a much longer time display extensive weathering. In a semiarid climate, like that of the Channeled Scabland, buried soils that developed on old flood deposits are rich in a calcium-carbonate precipitate – called caliche – that takes many thousands of years to accumulate into a hardpan.

Unfortunately, evidence of the earliest floods is limited to a few, isolated, widely distributed places in the region. Within the Channeled Scabland almost all the evidence for the much older floods is preserved in the Cheney-Palouse Scabland Tract. Perhaps, as more roads and excavations expose new outcrops of flood deposits, more evidence for old floods will be uncovered to help work out the chronology for

the earlier Ice Age floods. But the Rosetta stone for the timing and frequency of Ice Age megafloods lies off the Pacific coastline where a continuous record of flooding is likely preserved. Perhaps someday a carefully placed borehole will reveal the complete picture of 2 million years or more of Ice Age floods' history.

The earliest floods likely began to erode into Palouse loess, leaving behind teardrop-shaped islands known as "streamlined Palouse hills" upon the Channeled Scabland. Turbulence associated with high-energy floodwaters cut deeper and with each successive flood eventually breached the loess cover and eroded further into the underlying layers of basalt bedrock.

After being eroded by the floods, some of the sediment collected along the downstream ends of obstructions or up into backflooded valleys along flood paths or in some of the valleys downstream. Not all the eroded sediment came to rest in these basins, however. Most remained mixed with the turbid floodwater and were quickly flushed out to sea with each of the hundreds or more, short-lived, outburst-flood events.

As one might imagine, the windblown Palouse loess and Columbia River basalt are two very different types of material and eroded very differently. The fine-grained, semicohesive loess eroded grain by grain, primarily by abrasion. Furthermore, surfaces of Palouse loess are relatively smooth and do not induce strong erosive turbulence in the floodwaters. In contrast, basalt surfaces are rough, a situation that promotes turbulence when engulfed in racing floodwaters. And even though it is hardened rock, basalt is riddled with cracks that formed long ago as the lava cooled. Cracks, or fractures, created weak spots where floodwaters could easily pluck out basalt blocks en masse. So, even though basalt was rock, the floodwaters ended up eroding it faster and more easily than the relatively "soft" loess deposits that lay on top

Early Flood Evidence at the Marengo Railroad Cut An outcrop exposure amongst the streamlined Palouse hills at Marengo (pictured) is one of only a few places for evidence of very old floods going back one million years or more. Here is preserved the stratigraphy of an extremely old deposit of flood gravels, which lies at the base of the exposure - beneath a thick sequence of old soil horizons (caliche paleosols) developed in loess. Windblown loess within lower paleosols, above the flood gravel, retain a normal (N) over reversed (R) magnetic polarity locked in at the time of sediment deposition. The age of this world-wide magnetic reversal has been dated elsewhere at 780,000 years B.P. A radiometric (thermolumines-cence) age of 800,000 years B.P. from sediment below the magnetic reversal is consistent with the magnetic-polarity age. Since the flood gravels lie below the dated horizons, it follows that an Ice Age flood that occurred more than 800,000 years ago is preserved at this site (Bjornstad et al. 2001).

of the basalt. This explains the phenomenon of why "islands" of streamlined Palouse loess were able to withstand repeated attacks by the floods, even out in the middle of high-energy scabland flood channels.

Where the floods were especially powerful and eroded into basalt, they left behind a topography referred to by Bretz as "butte and basin," which characterizes much of the Channeled Scabland. In places, the floods scoured long, straight grooves in the basalt bedrock. As grooves expanded, the floods gouged out potholes that merged and coalesced into larger rock basins.

As erosion continued, vertical walls of basalt called cata-racts (now dry waterfalls) receded upvalley, leaving behind deep, steep-walled, coulees below the cataract cliffs. Scalloped, horseshoe-shaped cataracts, often dividing into two or three alcoves, receded upvalley for miles. Later floods might continue to widen the channels, cutting deeper into the basalt, creating multiple tiers of recessional cataracts and inner canyons.

After removal of the cover of Palouse loess, the many unique erosional landforms of the Channeled Scabland are as much the result of the Columbia River basalt bedrock as they are about the floods themselves. Multiple layered basalt

flows with regular, alternating fracture patterns between entablature and colonnade were conducive to the preferential plucking and rapid disintegration of the basalt flows, espe-cially along columnar zones. This promoted the rapid deep-ening and undercutting of basalt flows during flooding, leading to the formation of potholes, recessional cataracts, rock benches, blades, mesas and buttes. In contrast, none of these types of landforms developed in granitic or other rocks along the flood route. Therefore, it was the combination of the floods and the unique erosional behavior of the basalt that created the characteristic landforms of the Channeled Scabland.

Routing of Outburst Floods

During the last glaciation, which ended in eastern Washington ~13–14 thousand years ago, fingers of the Cordilleran Ice Sheet crept south across the northern border of the United States into Washington, Idaho and Montana. These glacial ice lobes blocked river drainages, including the Clark Fork near the Idaho-Montana border. The blockages created huge, ice-dammed lakes. The largest of these was glacial Lake

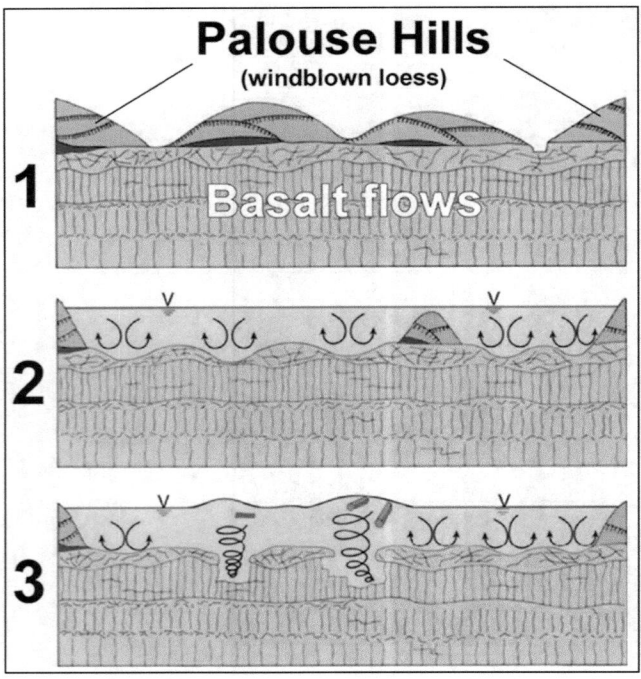

Palouse Hills
(windblown loess)

1 Basalt flows

2

3

Scabland Evolution Stage 1: Before Ice Age floods a thick blanket of windblown Palouse loess covered a stacked sequence of layered basalt lava flows. Stage 2: Early floods begin to erode the Palouse cover off the basalt bedrock leaving behind eroded, streamlined hills of loess. Stage 3: More erosion during subsequent floods strips away more Palouse loess while carving grooves, channels and rock basins deeper into the basalt bedrock. Modified after Baker (2009b).

Missoula, which extended 200 miles east, creating an inland sea of fresh, glacial meltwater equal to half the volume of Lake Michigan. The ice dam that created the lake was located over what is now Lake Pend Oreille in northern Idaho. Glacial Lake Missoula was up to 2000 feet deep and covered 3000 square miles of western Montana.

Flowing water naturally follows the path of least resistance and the Ice Age floods were no exception. After their escape from Lake Missoula, the floodwaters naturally followed low points in the landscape. The last floods from Lake Missoula raged first through Rathdrum Prairie, a mountain valley between what is now Sandpoint, Idaho and Spokane, Washington. Because Lake Columbia was already in place for most Missoula floods, floodwaters quickly finished the job of filling the Columbia valley behind the Okanogan Lobe ice dam. Upon reaching the brim of the valley, the floodwaters spilled over several low divides to the south and onto the area of today's Channeled Scabland. During most of the last glacial cycle the Okanogan lobe effectively stopped any Missoula floods from continuing down the Columbia valley west of Grand Coulee. Thus, flood after flood from Lake Missoula repeatedly swept through creating and eroding the Channeled Scabland. Evidence suggests that one or more Missoula floods early in the last glacial cycle continued west

all the way down the Columbia Valley, instead of across the Channeled Scabland at a time when the Okanogan Lobe had not yet fully extended across the Waterville Plateau.

After filling the Columbia Valley behind the Okanogan lobe, Missoula floodwaters spread across the Channeled Scabland and Mid-Columbia Basin via three main paths: (1) Grand Coulee, (2) Telford-Crab Creek and (3) Cheney-Palouse Scabland Tracts. A fourth route, Moses Coulee, also was used when the Okanogan lobe was not fully advanced. Combined, the floods' path across the scablands, from edge to edge, spreads out for 100 miles! Depending on the size of the flood, it may have occupied one or many of the scabland routes at the same time. Once the flood-channel network was established early in the Ice Age, later floods followed the same routes, sometimes perhaps with only minor changes.

Floodwaters that spilled over into the Cheney-Palouse tract headed almost straight south across the Palouse Slope, eroding the land down to basalt bedrock and stripping away up to hundreds of feet of Palouse loess. Much of this water drained into the Palouse and Snake rivers, while the remainder went west down Washtucna Coulee to join floodwaters coming from Grand Coulee and the Telford-Crab Creek tract.

Within hours, floodwaters from all three paths converged onto the central Columbia Basin near Pasco. But there was too much water for all to flow through the one and only outlet at Wallula Gap. The gap was too narrow and basalt ridges along the gap were too high for the floodwaters to cross over, so they backed up and created a huge temporary lake called Lake Lewis. For several days, this lake filled the mid-Columbia basin, backflooding up into the Yakima and Walla Walla valleys. In the Tri-Cities, only the crests of the higher ridges poked out above the lake that was up to 900 feet deep!

Recent computer models indicate Lake Missoula completely drained within four days or less of ice-dam breakup while the floodwaters took up to three weeks to find their way to the Pacific Ocean. (The delay in drainage was the result of temporary pooling behind several hydraulic constrictions starting at Wallula Gap and continuing through the narrow Columbia Gorge). After each flood the Cordilleran Ice Sheet continued to advance, and within a few years the ice dam began to re-block the Clark Fork River, initiating a new Lake Missoula.

The timing and frequency of climate change and glacial periods are well-preserved in sediment cores collected at depth in the Pacific Ocean. Here sediments have been laid down uninterrupted for millions of years and there is a continuous record of at least nine major glacial-interglacial cycles over the last 800,000 years. As many as several dozen of these glacial climatic cycles may exist going back to the start of the Ice Age, which began about 2.6 million years ago. Some geologists believe as many as 100 floods occurred during the waning stage of the last glacial cycle alone. If that's the case and if as many floods occurred during previous glacial cycles as the last, then the total number of floods for the entire Ice Age could easily add up to hundreds or even a thousand or more!

Another source for Ice Age flooding occurred ~18,000 years ago, near the peak of the last Ice Age, with a single flood from Utah's Lake Bonneville. Floodwater began to escape from the enclosed basin here at a low point for the lake at Red Rock Pass. At first, the lake overflowed slowly as it eroded into loose sediment of an alluvial fan complex shed off the surrounding mountains. Once started, however, erosional acceleration of the lake waters rapidly cut a channel through the alluvial dam. The ensuing flood lasted for about two weeks until floodwaters cut a 400-ft deep channel across the pass. Then the flood shut down upon encountering the resistant bedrock below. Altogether half the volume (~5200 km^3) of Lake Bonneville, flowing at up to one million m^3/sec, escaped through the pass.

Lastly, about 14,000 years ago, several hundred years after the last outburst flood from glacial Lake Missoula, there was one more flood. This flood occurred as the retreating Okanogan Lobe broke apart releasing the contents of glacial Lake Columbia. The Okanogan Lobe blocked the Columbia River near today's Grand Coulee Dam. With the ice dam created by the Okanogan lobe breached this last flood was restricted to the Columbia Valley, which totally bypassed the Channeled Scabland of eastern Washington. Unlike Lake Missoula, glacial Lake Columbia did not repeatedly fail on its own. Perhaps it was because Lake Columbia was a shallower lake (maximum 1500 ft deep) in contrast to Lake Missoula (more than 2000 ft deep). Perhaps the reduced buoyancy and pressure behind a shallower Lake Columbia might have kept pressurized water from destabilizing the ice dam.

Diagnostic Flood Features

Ice Age floods created an unusual collection of landforms produced by both erosion or deposition (Table 2). In general, the floods created mostly erosional features, like those in the Channeled Scabland, where floodwaters moved fast and furiously across the Palouse Slope. High flow rates were also generated, even in areas of relatively flat terrain, where floodwaters were forced to squeeze through narrower openings like Wallula Gap. It's a natural law, called the venturi effect, which makes the flow of fluids automatically speed up when forced through a bottleneck. (As an example, think of the difference in the speed of the water flowing through a bathtub drain versus the rest of the bathtub.)

Few features of the megafloods described below are common to all three flood sources (Lakes Missoula, Bonneville, and Columbia). Perhaps the only feature common to all three sources are giant flood bars covered with oversized boulders. More features, like ice-rafted erratics, are common to two of the three sources (Missoula and Columbia), since there was no ice dam associated with the Bonneville flood.

Erosional Features

Anastomosing Channels and Coulees (Missoula and Bonneville floods)

Flood channels are long, continuous, straight to broadly curving features eroded by short-lived outburst floods. A coulee is an elongated, box-shaped valley with steep or terraced walls and a wide flat bottom. Most coulees today are completely dry, or in a few cases, occupied by an "underfit" stream. An underfit stream is one that is greatly out of proportion with the size of the coulee it occupies. The bottoms of coulees are generally flat, except where they suddenly drop off over cascades or cataracts sometimes hundreds of feet tall. The Channeled Scabland is characterized by a network of channels and coulees that repeatedly divide and rejoin, forming an interconnected, braided-stream pattern, or anastamosis.

Deep, box-shaped coulees with steep walls and flat bottoms are characteristic of the southern Channeled Scabland where floodwaters incised into the basalt bedrock, especially below recessional cataracts. As floods receded, water levels would be confined to lower and lower channels. Continued flood erosion along a lower channel would often cut off higher channels, thus leaving them hanging. Later Ice Age floods were also smaller and therefore may have not occupied the older and higher flood channels but would truncate and cut them off at their lower ends.

Table 2 Diagnostic Ice-Age-Flood Features

Erosional Features
Interconnected networks of channels and coulees
Longitudinal grooves
Rock basins and potholes
Rock benches, mesas and buttes
Rock blades
Cliff overhangs and rock shelters
Faceted escarpments
Pinnacles and pillars
Divide crossings and spillover channels
Cascades, cataracts and plunge pools
Trenched spurs
Hogbacks
Residual, streamlined and scarped sedimentary hills
Abandoned spillways and hanging coulees
Giant eddy scars
Natural bridges
Strandlines
Exhumed ring craters
Depositional Features
Giant flood bars
Giant current ripples
Giant gravel mounds
Slackwater rhythmites (Touchet Beds)
Ice-rafted erratics, erratic clusters and bergmounds
Clastic dikes

© Springer Nature Switzerland AG 2021
B. N. Bjornstad, *Ice Age Floodscapes of the Pacific Northwest*, https://doi.org/10.1007/978-3-030-53043-3_2

Divide Crossings and Spillover Channels (Missoula and Bonneville floods)
A special type of channel associated with the Ice Age floods is the "spillover channel." These elevated, now-dry, channels developed where floodwater spilled over into adjacent valleys. If enough floodwater moved through the divide crossings, a flat-bottomed channel with planed-off, scarped sides often developed, giving the channel a trapezoid-shaped profile. Geologists use these spillover channels to identify the maximum heights of floods - useful for reconstructing the height and slope of the highest water surfaces during flooding. With this information, scientists using computer models have been able to confidently make estimates on the depth, speed and duration of the floods.

Longitudinal Grooves (Missoula floods)
Once the floods eroded away the unconsolidated, post-basalt sediment one of the earliest stages of erosion of the underlying basalt was formation of mostly straight, parallel, longitudinal grooves separated by low, flood-molded, bedrock ridges of basalt. Floodwaters eroded longitudinal grooves into the upper surface of basalt flows, mainly in the entablature portion. Grooves generally range from 100 to 200 feet apart, and about 10 feet deep.

Giant Longitudinal Grooves Near Dry Falls

Rock Basins and Potholes (Missoula and Bonneville floods)
As fast-moving floodwaters passed through scabland channels they gouged farther into the basalt, scouring out rock basins and augering deep holes into basalt. Like a powerful vacuum cleaner, floodwaters actually swept up all the loose material off the land surface, including huge columns of basalt, taking advantage of any weaknesses in the rock, such as fractures or rubbly basalt-flow bottoms. Deeper basins are often filled with water, where floods eroded holes below the present water table. Rock basins are often elongated and oriented in the direction of flood-water flow. Other scoured-out depressions in scablands include circular potholes, which were literally drilled out by violent, swirling vortices within the floodwater called kolks (Table 2).

Cliff Overhangs and Rock Shelters (Missoula floods)
Overhanging cliffs of basalt, some of which have served as rock shelters for indigenous Americans for thousands of years, are common in the steep walls of coulees, rock basins and potholes. The cliff overhangs formed as a result of differential erosion by Ice Age floodwaters, which locally

eroded away the more easily plucked, large columns of basalt in the colonnade and rubbly basalt toward the base of some lava flows. Thick layers of more cohesive overhanging entablature resisted these plucking forces to a greater degree.

Rock Benches, Mesas and Buttes (Missoula and Bonneville floods)

Flood-Excavated Rock Benches Along Moses Coulee

After carving channels into the basalt, the floods peeled away the once continuous basalt layers, one-by-one, away from the walls of coulees and rock basins. Since the uppermost basalt layers were attacked first, they are typically eroded farther away from the center of the coulee. The end result was a stair-stepped, tiered landscape of rock benches, as well as isolated mesas and buttes. Rock benches are flat areas, usually eroded out along the base of a flow, bordered by basalt cliffs of the next younger flow. They developed where a relatively weak colonnade, or occasional sediment layer, lay between more resistant layers of basalt entablature. Floodwaters preferentially eroded away the weak layers, undercutting the rock. Unstable, the cliff collapsed above the undercut, maintaining the vertical wall of basalt above the stronger cap rock on the bench below. Continued undercutting resulted in continued collapse and recession of the wall down to the underlying rock bench.

Mesas represent isolated, flood-eroded remnants of once-continuous volcanic-basalt flows. They are similar to rock benches except their upper surface is surrounded by cliffs on

all sides. Mesas are wider than they are tall, and their flat top is normally composed of more resistant basalt entablature. Buttes are like mesas except they are taller than they are wide. Over time, after repeated attack by floodwaters, rock benches may evolve into mesas, which may eventually be reduced to isolated buttes via erosion by multiple floods.

Rock Blades (Missoula floods)

Rock blades appear as elongated rocky buttes or mesas. They typically form as an erosional remnant between two recessional cataracts. Here a narrow rib of basalt bedrock may be left behind between two or more of the alcoves as the cataracts recede headward. During further flood erosion a rock blade may segregate into one or more isolated, elongated rocky buttes or mesas. Rock blades that are detached from the receding cataract that formed them are also called "goat islands." Bretz named them after the Goat Island at Niagara Falls, which he used as his model for Dry Falls and related features.

Pinnacles and Pillars (Missoula floods)

Occasionally, Ice Age floods scoured fractured basalt flows into tall, isolated pinnacles, or pillars (next page). Pinnacles and pillars are much taller than they are wide and have uneven tops, unlike mesas and buttes that usually have flat tops. Some of these pointed features may represent the eroded remnants of past mesas and buttes from which they evolved after repeated episodes of flooding. Pinnacles and pillars, like mesas and buttes, usually have a protective cap of more resistant entablature or flow-top breccia, leading to their preservation.

Cataract Cliffs and Plunge Pools (Missoula and Bonneville floods)

As floods raced across the Columbia Plateau, single to multi-alcoved recessional cataract canyons formed. Cataracts are characterized by tall, now-dry cliffs with horseshoe-shaped alcoves – some hundreds of feet high. At the bottoms of many cataracts lie deep, round plunge pools. Plunge pools are a special type of rock basin gouged out as floodwaters dropped vertically off the tall cataract cliffs. Today, plunge pools are often naturally filled with groundwater-seepage lakes. Below a cataract is an inner canyon that follows the path of the cataract recession. Cataracts, like rock benches, formed by more rapid erosion and undercutting of weaker basalt layers (colonnade) or interlayered sediments at the base of the cataract. In some places, such as the upper Grand Coulee, up to 30 miles of cataract recession occurred from undercutting at the base of the cataract as it steadily retreated up the valley with each subsequent flood. Similarly, a second recessional cataract is responsible for the Lower Grand Coulee canyon that retreated 15 miles from Soap Lake to present-day Dry Falls. Sometimes multiple cataracts are

Flood-resistant Basalt Pillars from Upper Grand Coulee

found along the length of a coulee. Initially, erosion scoured down to a weak layer in one of the upper basalt flows. Over time, erosion steps down to deeper weak layers in the basalt sequence, beginning a new recessional cataract while the original cataract has receded farther upvalley.

Trenched Spurs (Missoula and Bonneville floods)
Because of the tremendous force and momentum behind the floodwaters, they tended to flow like a fire hose, in straight lines or broad arcs. This is unlike normal rivers, which lazily flow along in sinuously curving meanders that often do not cover their valley bottoms, even during normal flood stages. In contrast, Ice Age floods were not constrained to valley bottoms and instantly overwhelmed the valleys they occupied, quickly overtopping the sides of the valleys and spilling into adjacent valleys. In some of these valleys, rapidly moving floodwaters overtopped spurs of basalt bedrock along the valley, which regular streams normally flowed around. An amazing feature called trenched spurs formed where floodwaters flowed straight up and over the rocky spurs, scouring one or more channels across the top.

Residual Streamlined and Scarped Palouse Hills (Missoula floods)
Streamlined and scarped Palouse hills developed where floods eroded and sculpted the blanket of windblown loess, that overlay basalt along flood paths. Unlike the underlying

basalt, loess is unconsolidated sediment – not rock. Fractures do not occur in these younger sedimentary deposits and therefore, no erosion occurred by plucking, the principal form of erosion in basalt. Instead, these landforms were molded via the much-slower, grain-by-grain abrasion by the floodwaters.

These hills consist of "islands" of loess that appear to float in a sea of "scrubbed" basalt scabland. Their teardrop shape, similar to an airfoil, consists of a prominent, steep prow on the upstream end and long, tapered tail on the downstream end. This shape, which was repeated over and over again in the formation of streamlined Palouse hills, is not a coincidence. During Ice Age flooding, the hills were streamlined so their length was usually about three times longer than their width; this hydrodynamic shape provided the least amount of drag to the floodwaters that formed them.

The sides of Palouse hills that lie along flood channelways were frequently planed off by the fast-moving, abrasive floodwaters. This resulted in the formation of faceted, planar surfaces called scarps that mark the upper level of flood erosion on these islands. Scarps are usually steeper and are aligned along margins of flood channels. They are distinctly steeper than the naturally rolling slopes of the hills themselves. Islands of Palouse loess, with steep sides and surrounded by scabland and with no evidence of more-gentle, pre-flood topography were likely completely overtopped by one or more Ice Age floods.

Faceted Face of Castle Rock (arrow)

Faceted Escarpments (Missoula and Bonneville floods)
Along the walls of high-energy flood channels, the immense power and speed of the floodwaters was enough to plane off and bevel the sides of coulees. Aligned scarps of loess facing the scablands are common along flood channels carved in Palouse loess but may also develop in basalt. Faceted escarpments are useful tools for determining the widths and depths of the floodwater, as well as determining changes in flood depth that occurred downstream.

Ringed Craters (Missoula floods)
In the area around Odessa, and a few other locations in the Channeled Scabland, are swarms of quasi-circular, ringed structures in basalt bedrock, eroded out by the Ice Age floods. These are variously named ring-dikes, basaltic ring structures or Odessa Craters. Most of these unusual structures, up to 1600 feet in diameter, have one or more concentric, raised rings with hollowed-out centers. Other ringed structures have central peaks surrounded by one or more ringed moats composed of near vertical dikes of basalt. The Odessa ringed craters formed in a flow (or flows) of the Roza basalt as the lava cooled about 15 million years ago. After that the Odessa Craters were blanketed in a cover of Palouse loess for eons. The buried ringed craters didn't appear until much later when the

basalt was exhumed by Ice Age floodwaters that etched and preferentially plucked out the looser and weaker rock within the ring structures.

While we know ring structures had to have formed as the basalt lava was being emplaced and started cooling, their exact origin is still under debate. Some ideas for their formation include: a sag-flowout model, in which the collapse of the hardened crust of lava into the molten interior squeezed up along concentric, dike-forming cracks, filling the sagged surface; earthquake tremors that shook the ground as the lava cooled; or even a meteorite that broke up before impact, coincidentally peppering the cooling lava flow with dozens of smaller impact features.

The most popular, recent explanation, however, is that the ring dikes are principally the result of violent steam explosions as the lava flowed over a wet surface. In this scenario pressurized steam is produced as the hot lava comes in contact and flashes with the water. The pressurized steam forces its way upward through the cooling lava flow along concentric vents, forming a pathway for still-molten lava in the flow interior to escape to the surface. Lava coming to the surface fills a depression created by the sagging flow interior. After solidification, ring dikes form where the lava once moved through concentric conduits to the surface. Millions of years later Ice Age floods moving over basalt removed the basalt

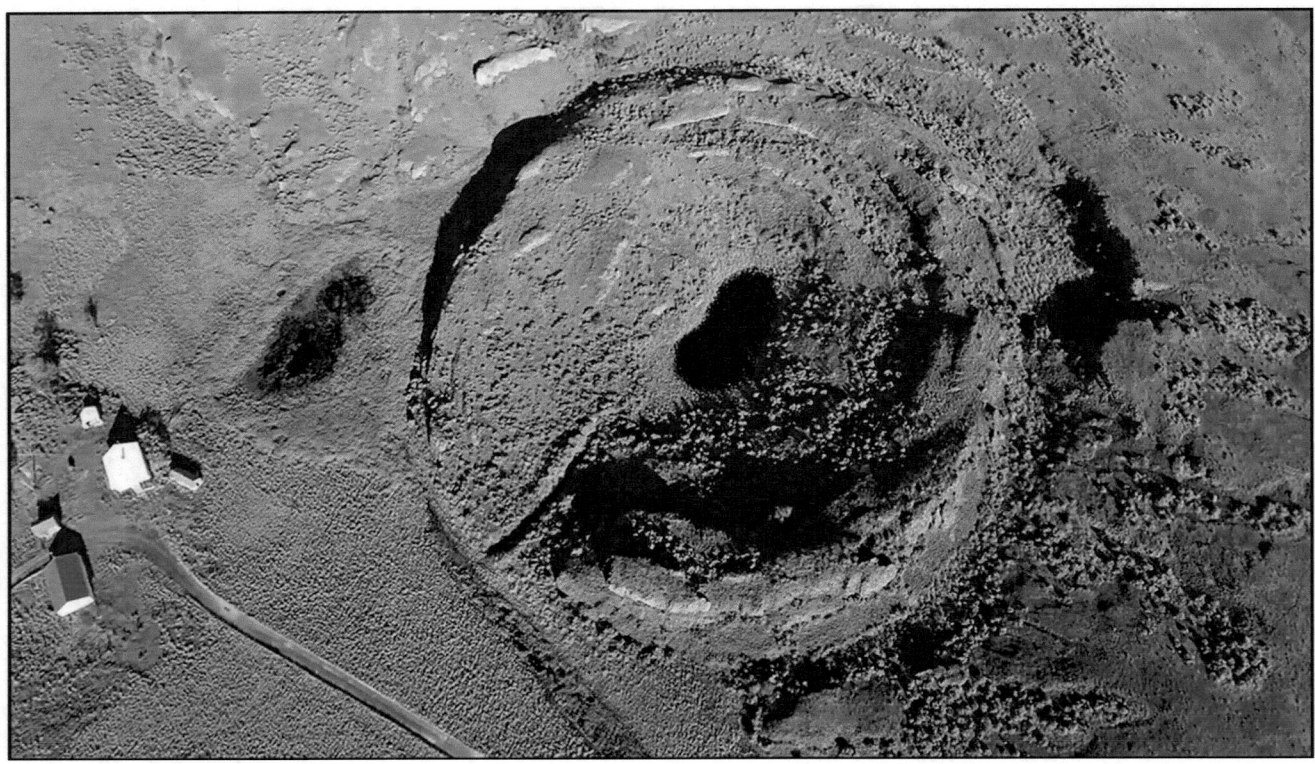

One of Dozens of Ringed Craters near Odessa WA

between the dikes more than the dikes themselves, leaving them to stand out in relief.

Depositional Features

The Ice Age floods locally scoured away up 250 ft of Palouse loess along with massive quantities of basalt bedrock across the Channeled Scabland. Because of all the suspended sediment, the ice-laden floodwaters may have resembled the color and consistency of a rich, frothy milkshake. And because of the high velocity of the floodwaters, which was maintained over much of the scablands, most sediment carried by the floods was quickly transported downstream, either into one of the many basins or eventually out to the Pacific Ocean. Locally within the Channeled Scabland, eroded gravel and sand did pile up into flood bars on the downstream sides of basalt obstructions or Palouse hills. Where the floods moved fastest, everything up to large boulders may have been suspended with the floodwater.

As floodwaters slowed the largest sediment grains settled out first, followed by smaller grains as the power and energy of floods continued to slacken. Coarse sediment also floated within icebergs, some of it ending up dropping out or rafting into quieter waters along the floods' route. Because of turbulence during flooding, much of the finer

sediment, such as sand, silt and clay, remained stirred up and suspended in the floodwater. The ultimate resting place for most of the fine-grained sediment, though, was on the continental shelf, many miles from where the Columbia River emptied into the Pacific Ocean (300–400 ft below modern sea level [see page 4]).

The floods naturally sorted the sediment moving along with the water, especially in areas where the floods expanded and slowed. At the base of the floodwaters' flow, gigantic boulders bounced and rolled along while a slurry of finer-grained clay, silt and sand remained suspended in the turbulent flow. In contrast, the top of the water column probably carried only fine sand-, silt- and clay-sized particles, along with any floating vegetation, animal carcasses and icebergs. As a result, coarse cobbles and boulders, except for those transported in floating icebergs, are generally absent at higher elevations.

Because of the high velocity and turbulence of floodwaters, often only coarse-grained sediments were deposited within the Channeled Scabland. Finer gravel and sand remained in suspension and were transported further downstream. Flood gravels are mostly composed of angular basalt clasts, which show a wide range of clast sizes. Gravel clasts are often coated with a thin layer of silt, as they settled through the turbid column of floodwater. These characteristics indicate the sediment did not travel very far before being quickly deposited and buried. In other places on the scab-lands, where the flood currents were less vigorous or at places higher along flood channels, only sand-sized sediment was transported and deposited. Sometimes coarse sand grades up into finer sand or silt-sized sediment grains form-ing sedimentary beds called "rhythmites." Most geologists agree that each rhythmite represents a deposit from a sepa-rate Ice Age flood event.

Giant, 300-ft tall Lake Sacajawea Bar (arrows) along the Snake River

Giant Flood Bars (Missoula, Bonneville and, Columbia floods)

Flood bars occur along most coulees and flood channelways, especially along inside bends and downstream of basalt buttes, mesas or upland plateaus, and below streamlined Palouse hills within, or along, flood channel ways. Elsewhere eddy bars formed where floodwaters swirled around the mouths of side canyons crossed by megafloods.

Giant Current Ripples (Missoula and Columbia floods)

Bretz did not have the luxury of aerial photographs until later in his career. As early as 1925, Bretz recognized that many flood bars displayed "a horizontal fluting and minor mounding on their side." By 1930 he had theorized they were giant current ripples, but it wasn't until years later, after aerial photography came into more common practice, that geologists began seeing these features at many places along the floods' path, thus adding credence to Bretz's "outrageous" flood hypothesis.

Geologists were originally clueless to their origin because of their huge scale and low relief, which made them difficult or impossible to detect at ground level. Just like ripples on a stream, giant current ripples are usually steeper on the down-stream side than the upstream side and provide an indication of flood-flow direction. Giant current ripples are among the features geologists have used to best estimate the speed and depth of the Ice Age floods. Based on the height and spacing of ripples, as well as the sizes of boulders transported, geolo-gist Vic Baker estimated floodwaters were traveling up to 50 miles per hour when the ripples were created and the water was up to 500 feet deep. More recent estimates have placed a maximum speed of the floods approaching 80 miles per hour. The sizes of giant current ripples generally increase with depth and velocity of the floodwaters that produced them.

Slackwater Rhythmites (Missoula floods)

Rhythmites are composed of graded, sedimentary beds with each bed representing a separate Missoula flood. In the Columbia Valley north of the Channeled Scabland sand beds at the bottoms of rhythmites represent rapid (few hours or days) deposition associated with a Missoula flood invading glacial Lake Columbia that persisted between many of the Missoula floods. The sand bed transitions upward into up to dozens of annual fine layers of silt and clay, called varves. Varves were laid down very slowly, over a span of dozens of years or more between flood events, each varve couplet representing a single year of deposition like tree rings, within Lake Columbia. Rhythmites may also occur in the Channeled Scabland but without the intervening varved lake beds; varves are absent since no permanent lakes existed there – all the flood flow was overland across a surface not covered by water between flood events.

Slackwater Rhythmites of Burlingame Canyon

Ice-Rafted Erratics (Missoula and Columbia floods)

Icebergs, from the sudden breakup of the ice dams, were carried along with the floodwaters. Boulders encased within ice hitchhiked their way to the Channeled Scabland and beyond via floating icebergs. Most of the boulders are composed of angular, exotic granitic and metasedimentary rocks, identical to the types of rocks observed today in the vicinity of the ice dam for Glacial Lake Missoula. Ice-rafted erratics are relatively scarce within the Channeled Scabland itself, however, since the deep and fast-moving floodwaters kept the icebergs constantly moving. This changed downstream, however, where the water slowed in several slackwater basins (e.g., Pasco Basin and Willamette Valley). This is where most icebergs grounded, leaving behind their payload of erratics. Occasionally, however, icebergs did become grounded, or hung up in eddies, or other traps along route through the scablands.

In the Pasco Basin, almost all erratics consist of rocks very different from the dark-colored Columbia River basalt, the only local rock type. About 75 percent of erratics are light-colored granitic rocks. The remainder are mostly metasedimentary rocks of the Belt Supergroup composed of mostly argillite and quartzite. Belt rocks are extremely old (1.4 billion to 1.6 billion years old!) that, not coincidentally, came from the vicinity of the ice dam for glacial Lake Missoula.

Glacial Lake Missoula Floods

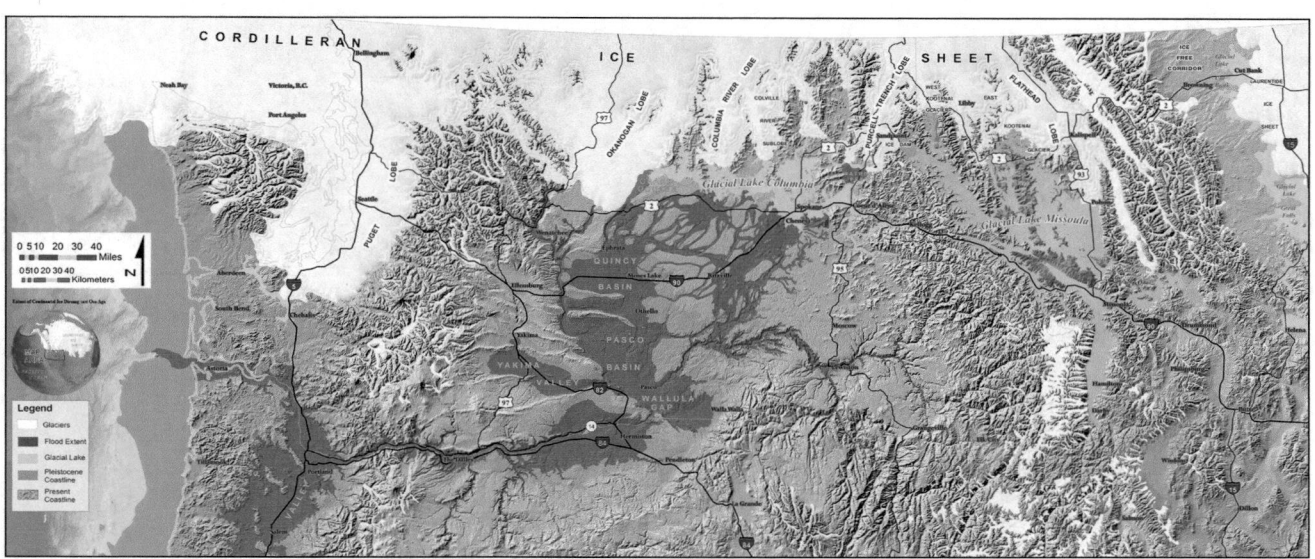

Lake Missoula Flood Map. Produced by Eastern Washington University in cooperation with the Ice Age Floods Institute.

Glacial Lake Missoula

Ice Age megafloods from glacial Lake Missoula occurred dozens of times when the Purcell Trench Ice Lobe repeatedly failed and released the contents of Lake Missoula. At the same time, glacial Lake Columbia, another huge lake downstream, was dammed behind another ice lobe (Okanogan) that per-sisted in place during most Missoula floods. Blocked by the Okanogan Lobe, Missoula floods instantly overfilled Lake Columbia before spilling south into the upper reaches of the Channeled Scabland. During the largest floods Lake Missoula emptied in just a few days but took up to three weeks to com-pletely drain to the Pacific as floodwaters temporarily ponded behind hydraulic constrictions along the way.

B. N. Bjornstad, *Ice Age Floodscapes of the Pacific Northwest*, https://doi.org/10.1007/978-3-030-53043-3_3

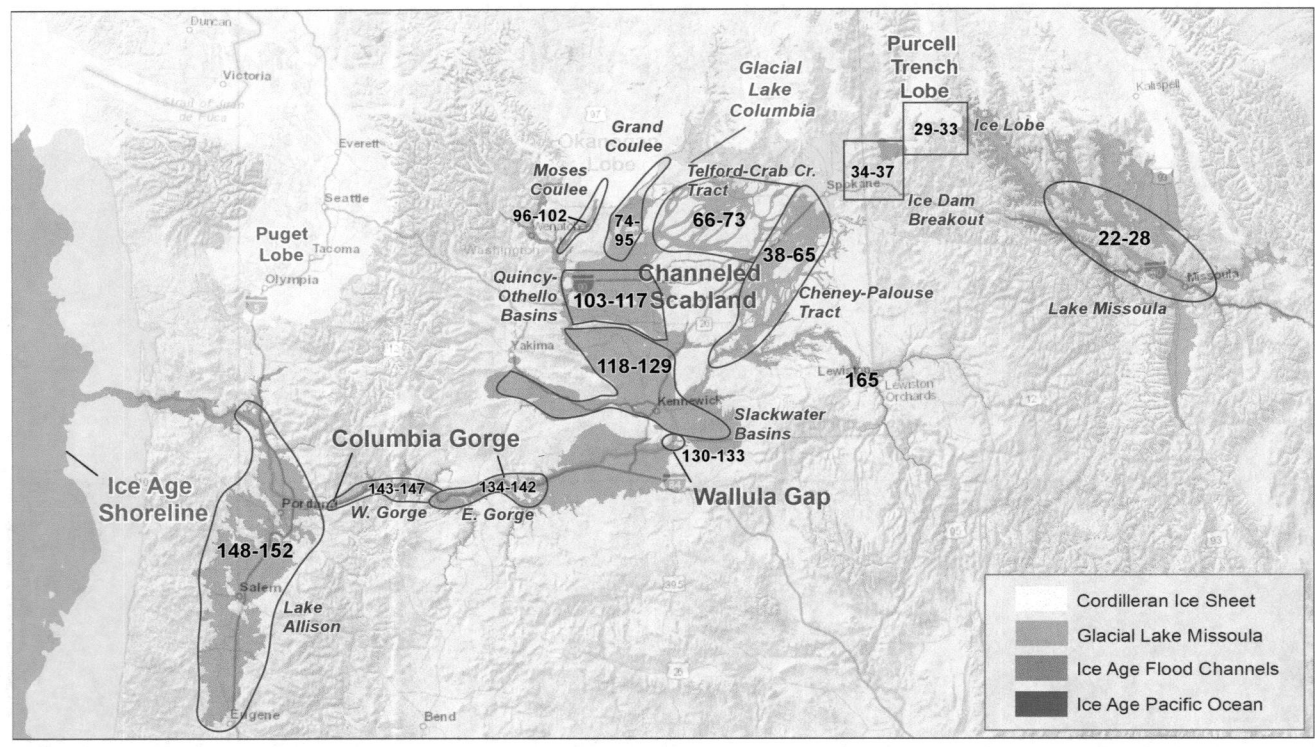

Index map for the Missoula Floods Numbers correspond to pages where flood features are located within the various flood-impacted regions. Base map courtesy of Washington State Parks and Recreation Commission.

Glacial Lake Missoula When glacial Lake Missoula was dammed by the Purcell Trench ice lobe it created a 2000-ft deep lake containing up to 600 cubic miles of glacial meltwater extending for 200 miles into western Montana. Numbers correspond to book pages herein.

Lake Missoula Maximum Strandlines disappear near the top of Mount Jumbo at ~4200 ft elevation indicating that glacial Lake Missoula reached a maximum depth of 2000 ft. After breakup of the ice dam Lake Missoula drained in only a few days – heading west toward the Idaho panhandle.

Strandlines of Glacial Lake Missoula Like giant bathtub rings, former lake levels of glacial Lake Missoula are marked by strandlines. Strandlines are especially well developed on west-facing slopes — where the fetch of predominantly west winds pushed crashing waves onto former shorelines of Lake Missoula. Mounts Jumbo (above) and Sentinel (below) rise directly above Missoula, Montana.

"At many places on the untimbered sides of basins up to an altitude of 4150 feet the horizontal markings of closely spaced lake beaches are plainly visible from moderate distances." Pardee (1942)

Camas Prairie Giant Current Ripples As Lake Missoula rapidly drained, water from the lake margins raced inward to fill the void. In the distance the speed of the floodwaters quickened temporarily as floodwaters flowed toward the camera through Markle Pass and Wills Creek Pass. Notice how the orientation of giant current ripples is markedly different on either side of SR 382 coming from the two different passes. The maximum stand of Lake Missoula rose 400 ft over the hilltop at center (arrow) while the passes on either side were 1000 ft below the highest stand! Giant ripples are up to 50 ft tall and 250 ft apart. Looking north.

Markle Pass Spill Through Looking north into the Little Bitterroot Valley through the Markle Pass floodwater-escape route. Floodwater drained through the pass toward the viewer. Erosional scouring within the pass quickly transitioned to deposition of coarse flood sediment into giant ripples and a massive, tongue-shaped delta bar (outline) just downstream of the pass all indicating a massive flow of the water was to the south (arrow).

"… the form of a bar about 50 acres in extent built out to a height of 150 feet over the slope descending to the basin."
Pardee (1942)

Wills Creek Pass Ripples   Above: An abandoned homestead is dwarfed by giant current ripples clearly coming from the direction of Wills Creek Pass. Flow direction, here toward the camera, is indicated by ripple asymmetry (steeper side faces downstream). Notice erosional scouring through the pass (arrow). Looking upstream to the north. Below: Imagine the destructive power and energy of water flowing through Wills Creek Pass to create these gargantuan ripples. Abandoned homestead and trees are the same in both images. Looking southeast.

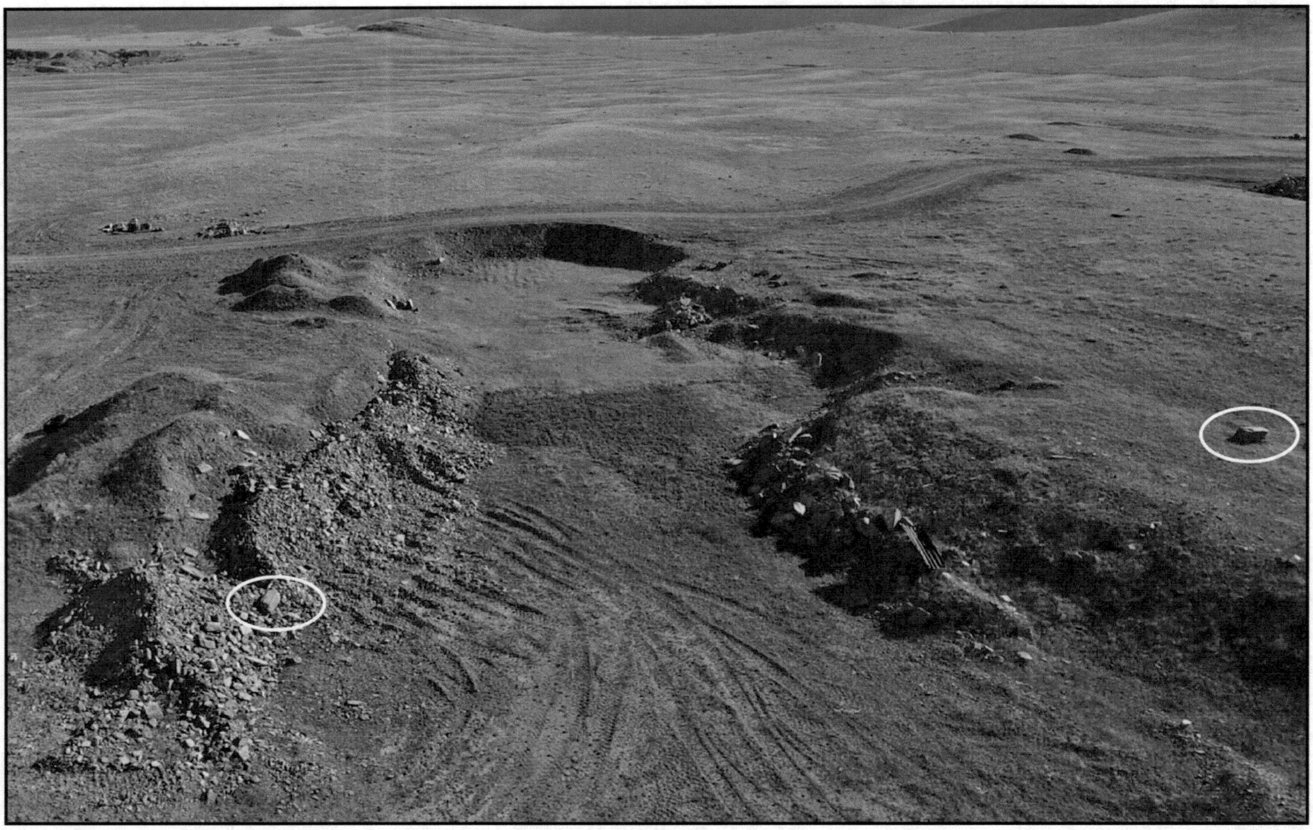

Powered-Up Flood Currents on Camas Prairie A quarry operation exposes the interior of a field of giant current ripples coming from Markle Pass. Glacial meltwater in Lake Missoula drained very rapidly, judging by the enormous size of argillite boulders (circled) coming from the mega-ripples. Flood currents were especially strong here because of floodwaters that were temporarily held back upstream in the Little Bitterroot Basin behind hydraulic constrictions at Markle and Wills Creek Passes.

High Eddy Deposit (Stout's Bar) along Flathead River   Downstream of Camas Prairie floodwaters entered the Flathead River Valley before eventually joining the Clark Fork River. Along the way flood-transported sediment accumulated in side gulches like this elevated eddy bar along the north side of the Perma Narrows. Notice how the height of the eddy bar (arrow) decreases away from the Flathead valley – the result of sudden Lake Missoula drainage and not a delta bar that prograded into former Lake Missoula.

"The high eddy deposits are characterized by a steep front slope, a rounded top, and a level or gentle back slope as illustrated by the deposit." Pardee (1942)

Eddy Narrows Constriction   Almost the entire volume (75%) of glacial Lake Missoula was forced to drain through this single opening, less than a mile wide and 10 miles long, at Eddy Narrows where the Clark Fork River flows today. The increased speed of the floodwaters through this narrow passage neatly cleaned off the canyon walls beautifully exposing hundreds of feet of 1.5-billion-year-old, metasedimentary Belt rocks (mostly argillite). Before draining, the largest Lake Missoula would have extended to near the top of this canyon. Using a flood depth of 1000 ft and an average cross-sectional area of 5.9 million square feet J.T. Pardee (1942) estimated a maximum flow rate for floodwater through the constriction at 385 million ft³/sec. More recently, this estimate has increased almost three fold (slightly over one billion ft³/sec) (O'Connor et al. 2020).

"At its high stage the lake is roughly estimated to have held more than 500 cubic miles of water of which nearly three-fourths was stored above a constricted part of the Clark Fork Valley called the Eddy Narrows" Pardee (1942)

Purcell Lobe Ice Dam

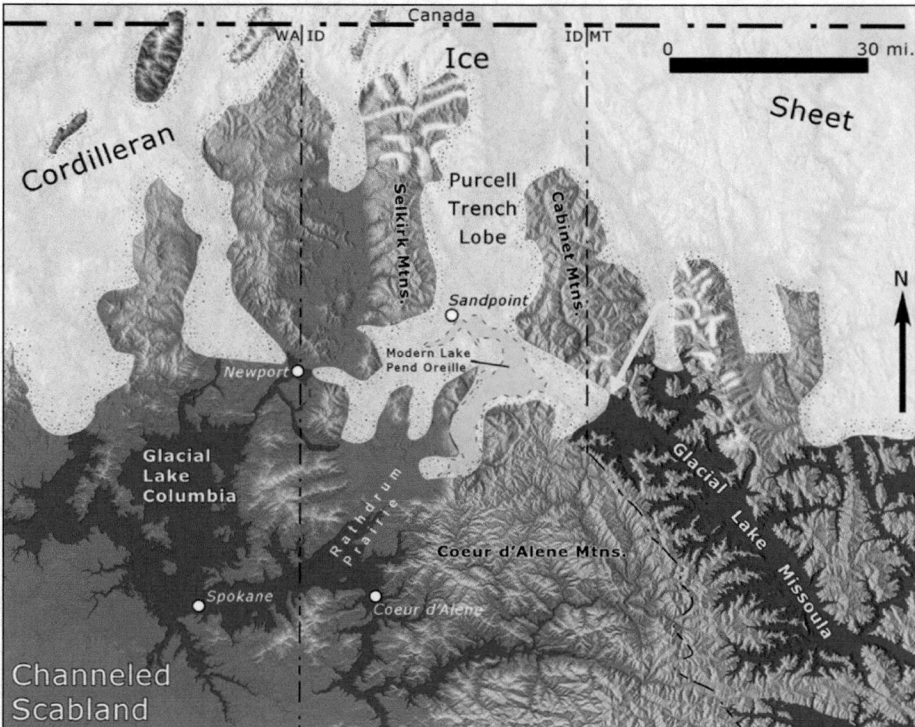

Lake Missoula Ice Dam The creation of the ice dam for Lake Missoula occurred here when the 3000-ft-thick Purcell Trench Ice Lobe crept UPVALLEY to block the Clark Fork River near the Idaho/Montana border. Completely sealed off, meltwater in the lake filled the Clark Fork valley up to 2000 ft deep. Upper image is a computer-modified aerial photo looking west at leading edge of the former ice dam taken above the Clark Fork valley by pilot Dave Bennett (www.glaciallakemissoulafloods.com). Arrows point to an approximately common point in the two images.

Ice-Dam Limit Near Cabinet Gorge Dam   Today, the Cabinet Gorge Dam lies near the eastern boundary for the Purcell Trench Lobe and ice dam for Lake Missoula. Looking east along the Clark Fork River from Idaho into Montana.

Belt Rocks and Ice-Rafted Erratics   Above is a quarry of extremely old rocks belonging to the Belt Supergroup – metasedimentary rocks composed of mostly quartzites and banded, slaty argillites. These distinctly layered rocks, thousands of feet thick, were laid down in ancient rivers and playas ~1.5 Billion years ago! Today they make up much of the bedrock beneath former glacial Lake Missoula and its ice dam. The blocky and colorful (red, purple, green, gray, brown) nature of the Belt rocks is apparent in the constructed rock wall above. Due to their extreme age and extended period of deep-burial regional metamorphism, Belt rocks are extremely hard, durable and resistant to weathering and decomposition. For this reason they are common as ice-rafted erratics in areas downstream of the Missoula floods.

Rattlesnake Mountain Argillite Hundreds of miles away an erratic boulder of blocky and banded argillite rests on Rattlesnake Mountain, located in south-central Washington – far from the nearest outcrop exposure of Belt rocks from which it is derived. The only rock native to this area is Columbia River basalt. Therefore, this boulder, and many others like it, could have only rafted here in icebergs after the breakup of one of the many ice dams that held back glacial Lake Missoula. Based on a survey of over 2000 ice-rafted erratics on Rattlesnake Mountain about 10–20% consist of distinctive Belt rocks composed of mostly argillite and quartzite.

Paradise Under Ice It's hard to imagine these scenes where the Clark Fork River empties into the head of Lake Pend Oreille were once about midway along the ~30-mile long and 3000-ft--thick ice dam for glacial Lake Missoula. Upper image looks west onto Lake Pend Oreille; lower image looks east from the head of the lake into the Clark Fork River valley.

Ice Dam Breakout

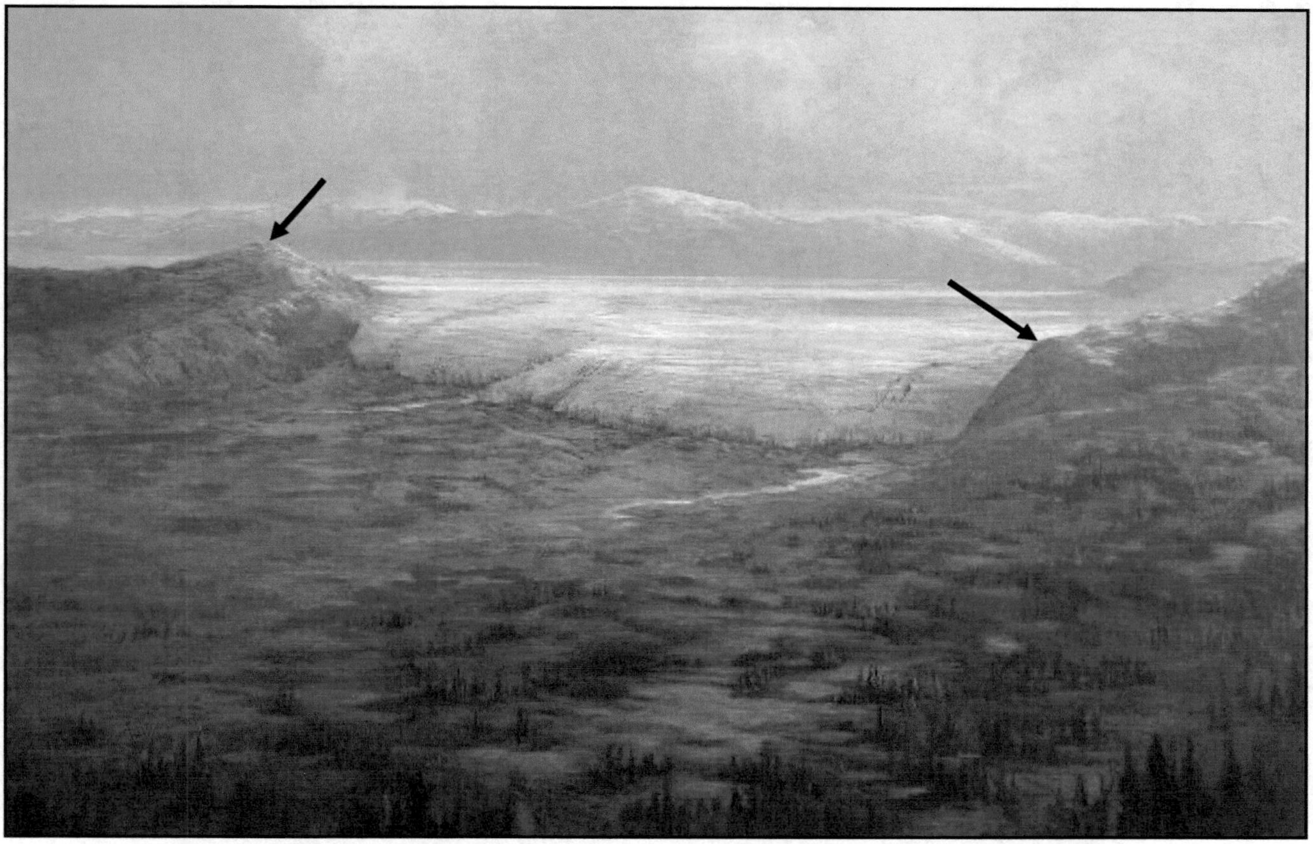

Ice Dam Breakout During the Ice Age the ice dam for glacial Lake Missoula extended up to 30 miles from the Idaho-Montana border to the lower end of modern Lake Pend Oreille (above). On top is a probable scene before a megaflood outburst illustrated by artist Stev Ominski. For reference, arrows point to the same two points in both images. Left arrow points to Cape Horn Peak; right arrow points to a break in slope where long-term abrasion by the Purcell Trench glacier over-steepened the valley wall. Peaceful Bayview appears to the left and Farragut State Park is in foreground.

Glacial Terminus Bayview is nestled into one arm of the former glacier that once replaced Lake Pend Oreille. Arrows point to the over-steepened valley wall created by glacial scouring near the terminus of the Pend Oreille Ice Lobe.

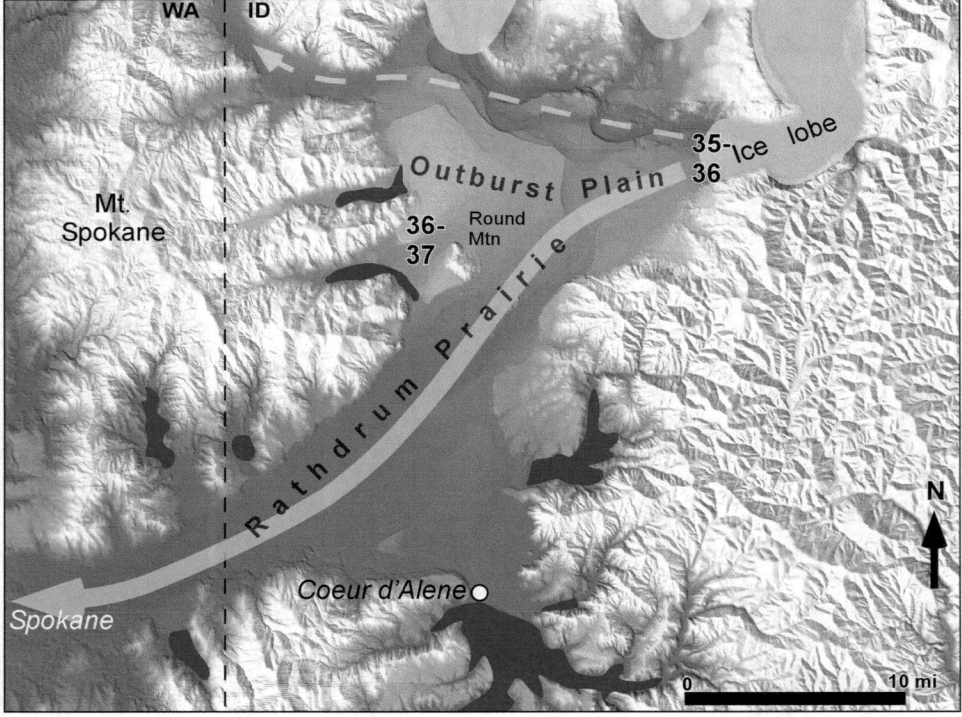

Glacial Lake Missoula Outbursts Beyond the breakout zone for Missoula floods lies the Rathdrum Prairie flood-outburst plain. Buildup from hundreds of feet of flood–deposited sediment onto Rathdrum Prairie blocked local tributary streams and rivers creating a series of debris-dammed lakes (dark blue) along the margins of the prairie. Depending on the location of ice lobes some of the floodwaters also went west (dashed), although most of the floodwater escaped via Rathdrum Prairie. Notice the 250-ft tall, triangular-shaped, pendant flood bar just downstream of Round Mountain. Numbers correspond to book pages herein.

Lowering Snout of Glacial Ice Dam   The lower end of the ice dam for Lake Missoula extended all the way to the south end of modern-day Lake Pend Oreille at Bayview and Farragut State Park. Today, the former terminus of the glacial ice dam is apparent from the planar, sheared-off slope above stunning Idlewilde Bay (arrow). On the opposite side of the valley the ice dam approached the top of Cape Horn Peak at upper left.

Rathdrum Prairie Outburst Plain   Above: a broad, almost-flat megaflood-outburst plain extends for dozens of miles across Rathdrum Prairie in the distance. Hundreds of feet of bouldery megaflood deposits lie buried beneath the plain. Due to the coarseness of the sediments the underlying Rathdrum Prairie is free of any rivers or streams since all water quickly percolates downward to feed the underlying, prolific, Spokane Valley-Rathdrum Prairie Aquifer. Arrows point to the once-descending terminus of the Pend Oreille lobe glacier. Looking south. Below: western plain of Rathdrum Prairie. Debris-dammed Spirit Lake at upper right. The largest outburst floods rose to the summit of Round Mountain (arrow).

Debris-Dammed Twin Lakes on West Side of Rathdrum Prairie The buildup of coarse megaflood deposits from repeated outburst floods raised the level of Rathdrum Prairie leading to the blockage of eight, pre-existing tributary streams (see page 34). This resulted in the formation of a series of debris-dammed lakes around the margins of Rathdrum Prairie.

Floods' Incursion

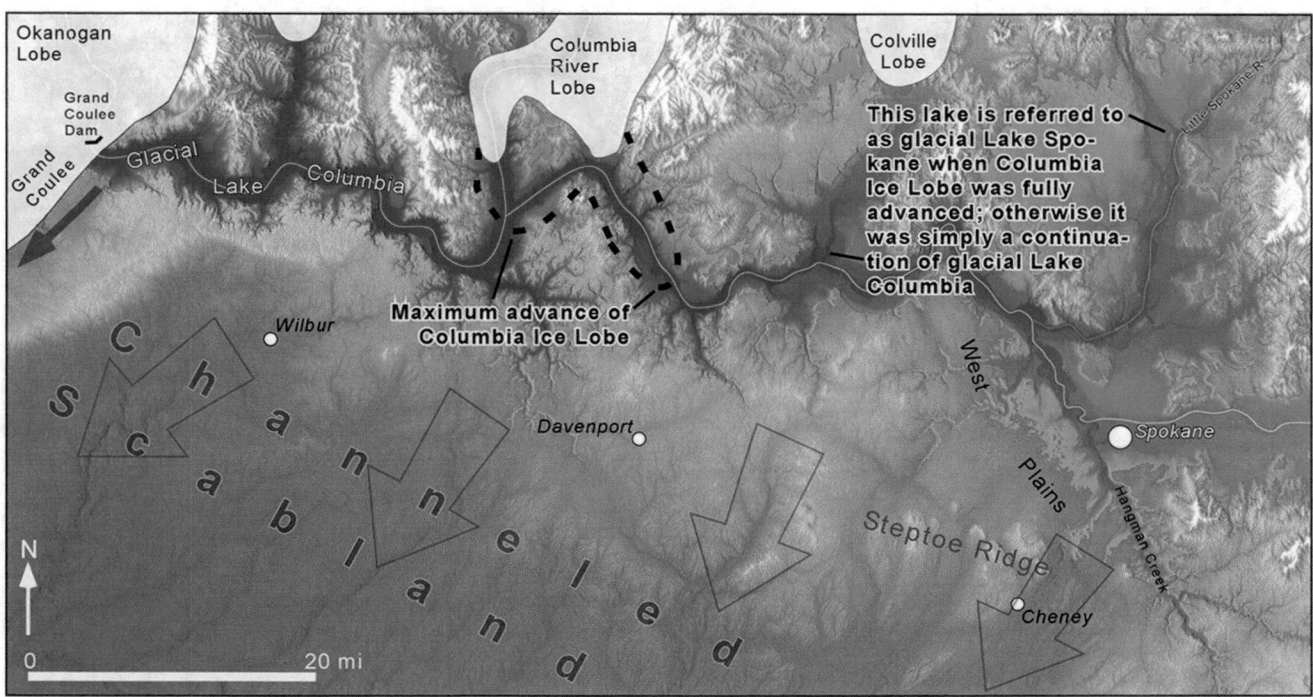

Glacial Lake Columbia Incursion During most Missoula floods another ice dam (Okanogan Lobe) existed downstream in north-central Washington. Earlier in the ice age glacial Lake Columbia, which once rose to ~2300 ft elevation, developed behind the Okanogan ice lobe, filled the Columbia and Spokane valleys and at times backed up all the way to Idaho. During this time, with the Okanogan ice lobe blocking the Columbia River valley, all Missoula floods were diverted south across the Channeled Scabland (arrows). It's also possible that the Columbia River lobe also advanced across both the Columbia and Spokane valleys to produce a separate lake (glacial Lake Spokane). When, and if, this happened outburst floodwaters from Lake Missoula would have all been diverted solely down the Cheney-Palouse Tract by way of the Steptoe Ridge.

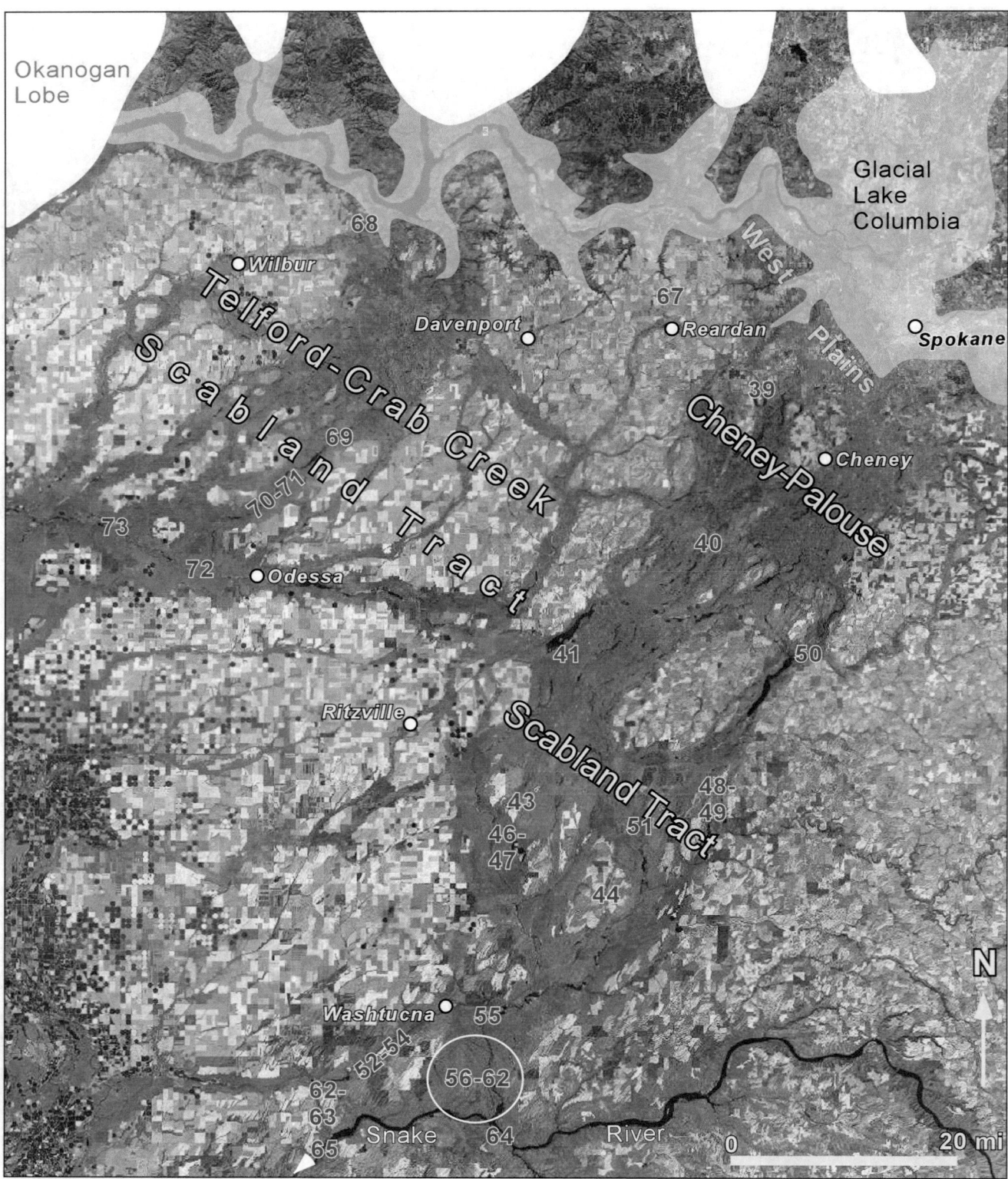

Index Map for the Eastern Channeled Scabland   The tortuous paths of the floods are easily discernable from space where the blanket of fine, wind-blown Palouse loess was stripped away down to the dark and bare basalt bedrock below—producing the "Channeled Scabland". The name "scabland" came into use by farmers and others who first settled the region. For they quickly realized that the barren and rocky areas swept clean of sediment (i.e., scabland) were not as productive as other nearby areas covered with Palouse loess. The southwestward path of megafloods is clearly visible with braided flood channels starting at the upper ends of the Cheney-Palouse and Telford-Crab Creek Tracts. Numbers correspond to flood features on pages described herein.

"The physiographic expression of the region is without parallel; it is unique, this channeled scabland of the Columbia Plateau." Bretz (1928a)

Cheney-Palouse Tract

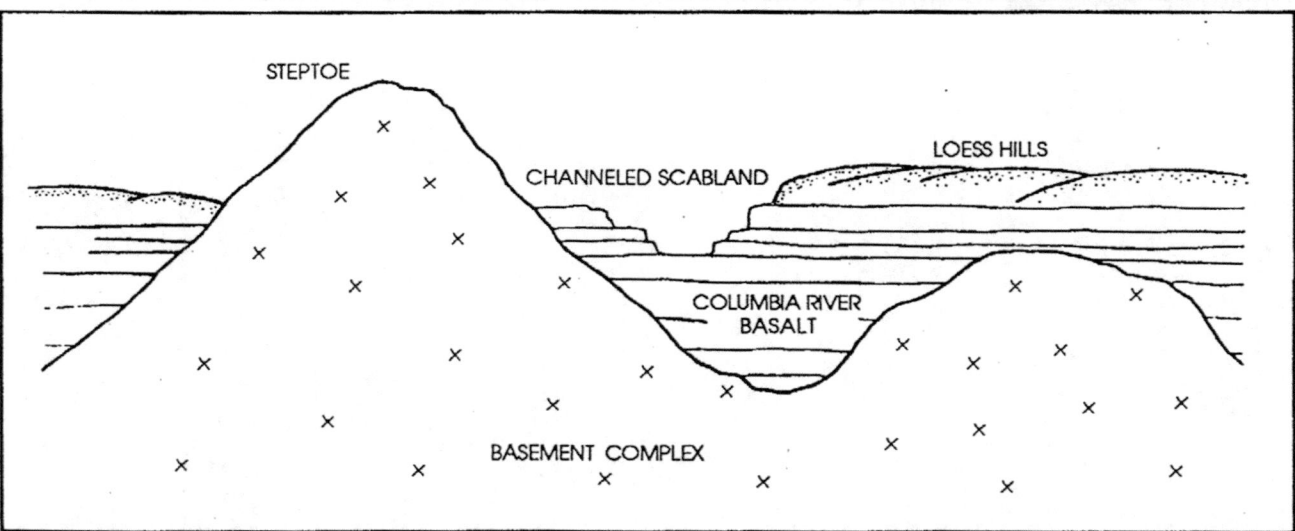

Medical Lakes Scour Channels   Above are the two Medical Lakes scoured out by the megafloods. Floodwaters entering the head of the Cheney-Palouse Tract scoured out these channels as they crossed over the Steptoe-Ridge spillover complex. The basalt flows here are thin to non-existent where they lapped up onto the Precambrian basement rock. This allowed megafloods to erode more deeply along the discontinuity. Looking southeast into upper reaches of the Cheney-Palouse Scabland Tract.

"Four heads northwest of Cheney contain nine named lakes in rock basins and multitude of rock-basined swamps and smaller lakes. Five of these lakes lie essentially on the plateau divide." Bretz, Smith, and Neff (1956)

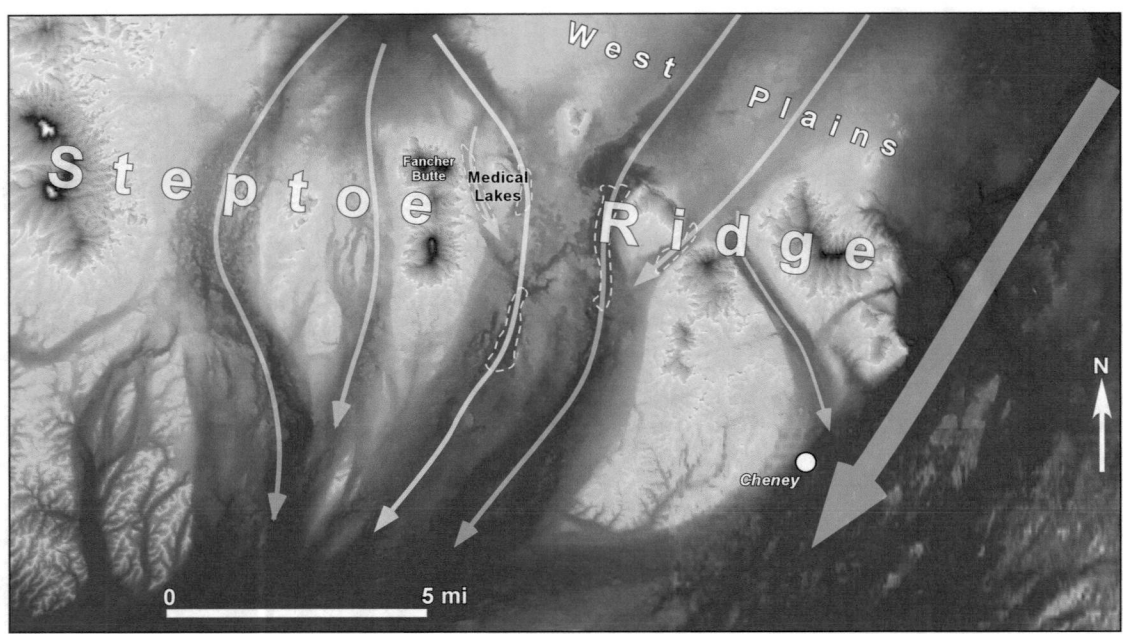

Steptoe Ridge Spillover Complex After invading glacial Lake Columbia, Missoula floodwaters spilled south into the head of the Cheney-Palouse Tract (see page 38). A series of semi-parallel spillover channels were gouged out as floodwaters eroded into a ridge of "steptoes" – older, pre-basalt basement rock that protrude above the much younger basalt. Modern lakes are outlined in yellow. Everything in blue was underwater during the largest Missoula flood. (~2500 ft elev.).

Hog Lake Falls Hog Canyon Creek, near the head of the Cheney-Palouse Tract, drops over a cataract cliff into megaflood-carved Hog Lake. Looking northeast, into the direction of Steptoe Ridge and the invading floodwaters.

Scablands Near Sprague Lake Megafloods totally stripped away the covering of Palouse loess down to bare basalt bedrock here at the upper end of the Cheney-Palouse Tract. During megafloods, some coarse sediment was deposited onto a flood bar covered with giant current ripples (between arrows) along Sprague Lake (above). Below: classic butte-and-basin scabland.

"No one with an eye for landforms can cross eastern Washington in daylight without encountering and being impressed by the 'scabland.' ... The region is unique: let the observer take the wings of the morning to the uttermost parts of earth: he will nowhere find its likeness." Bretz (1928a)

Rolling Palouse Hills   In between Scabland channels of the Cheney-Palouse Tract are islands of gently rolling hills of Palouse loess, up to 250 ft thick, that lie above the maximum flood level. Today these hills of fertile, wind-deposited loess support a booming agriculture – mostly dryland wheat (top) but also other crops like canola (below). Notice the mature, gentle, tree-like dendritic drainage pattern characteristic of the Palouse hills. Hidden below these loess hills lies relatively flat layers of hard basalt bedrock. See also page 7.

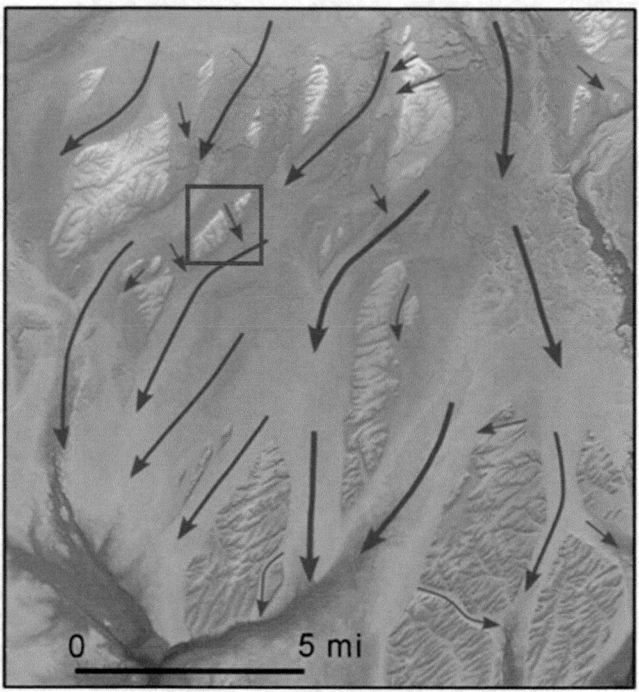

Flood-Streamlined Loess Islands   Before the megafloods this entire area was blanketed with hundreds of feet of windblown loess. Subsequently, Ice Age megafloods created the Channeled Scabland – a braided network of channels eroded into the barren basalt bedrock, separated by islands of fertile Palouse loess that support today's burgeoning wheat industry. Arrows indicate flood-flow direction toward the south. Notice streamlining of the Palouse hills by the floodwaters with a prominant "prow" pointing upstream. Red box outlines same island as that above.

"Literally hundreds of isolated groups of maturely eroded hills of loess stand in the scablands. Their gentle interior slopes are identical with those far from scabland tracts. But their marginal slopes, descending to the scablands, commonly are very steep." Bretz (1923)

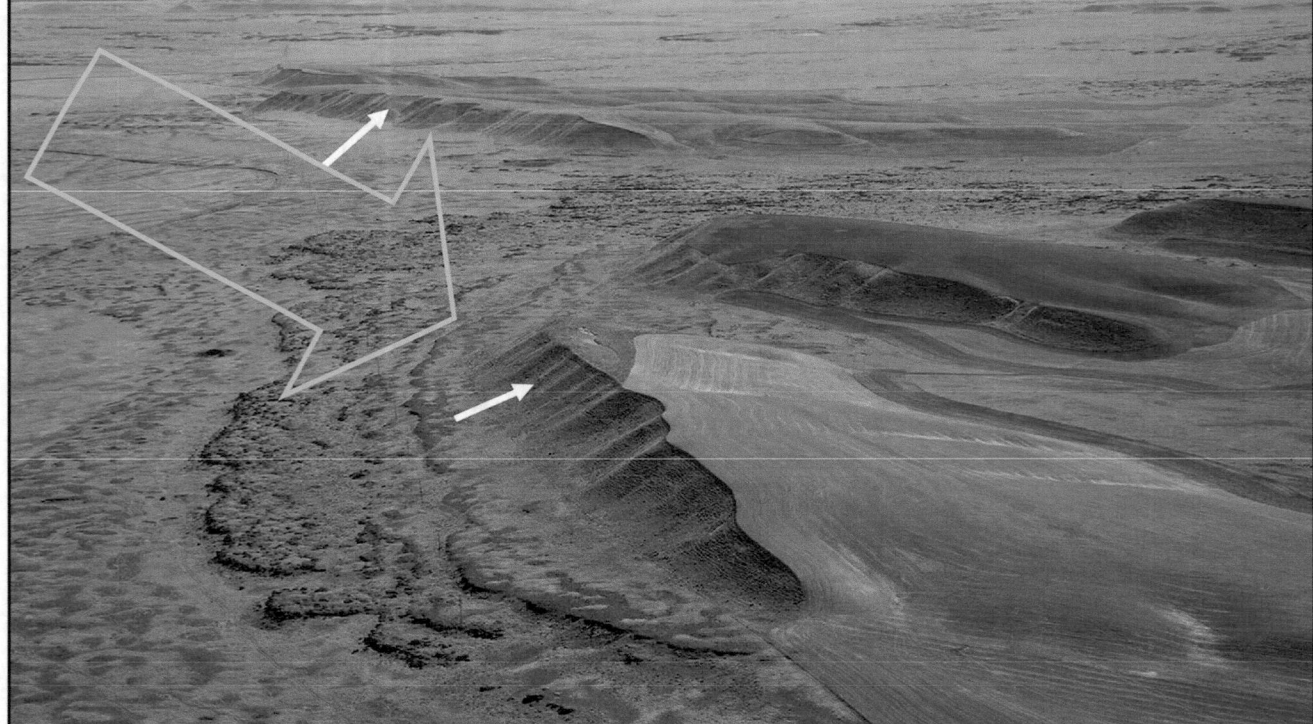

Palouse Escarpments Long lines of over-steepened, faceted escarpments (white arrows) developed along the sides of some flood channels, trimmed off by erosion by fast-moving floodwater. Block arrows indicate megaflood flow direction.

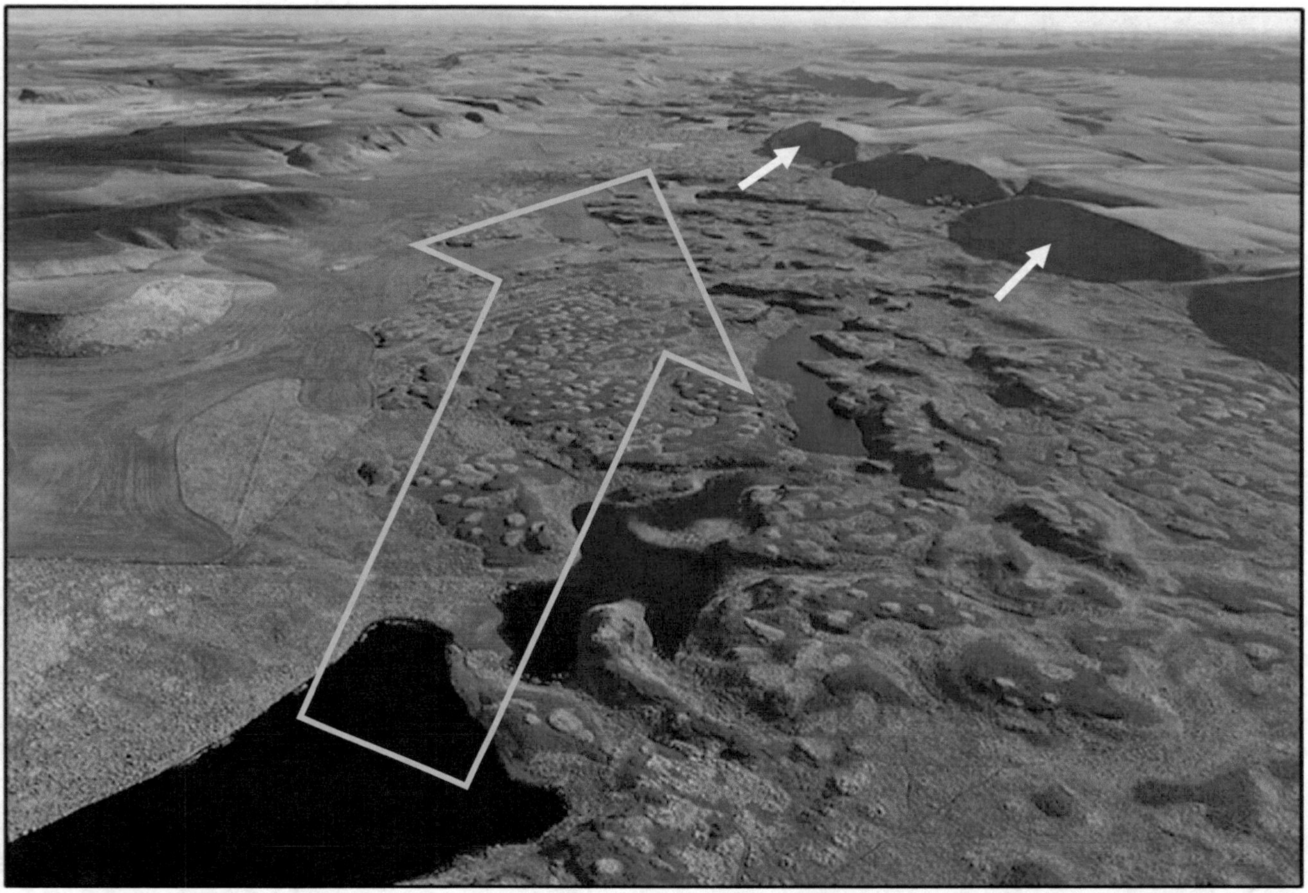

"It seems clear that they are but remnants of a once continuous cover of the basalt, and that the scablands have resulted from removal of the Palouse Hills by erosion by some unusual way. The basalt of the scablands is the firm and resistant foundation on which the hills stand." Bretz (1923)

Solitary Palouse Islands   Flood-streamlined Coyote Butte (below) stands alone. Megafloods almost succeeded in stripping away all loess off the basalt bedrock here, except for this single, streamlined Palouse island. The surrounding erosional escarpment goes all the way to the summit—indicating floodwaters once overtopped the island. Within the Cheney-Palouse Scabland Tract near Benge, WA. Floodwaters overran the island above carving a spillover channel across the top (small arrow).

Cow Creek Scabland Erratic Local landowner is dwarfed by a massive boulder of exotic granodiorite. This ice-rafted erratic from two vantage points rests along a flood channel in the Cow Creek drainage within the western Cheney-Palouse Tract near Marengo. Ice-rafted erratics are rarely observed along high-energy scabland channels like this one, but due to its humongous size this behemoth was apparently too large to carry away – even for the extremely powerful floodwaters that flowed through here. More recently, industrious pioneers removed rectangular slabs off the boulder—apparently for use in local house foundations.

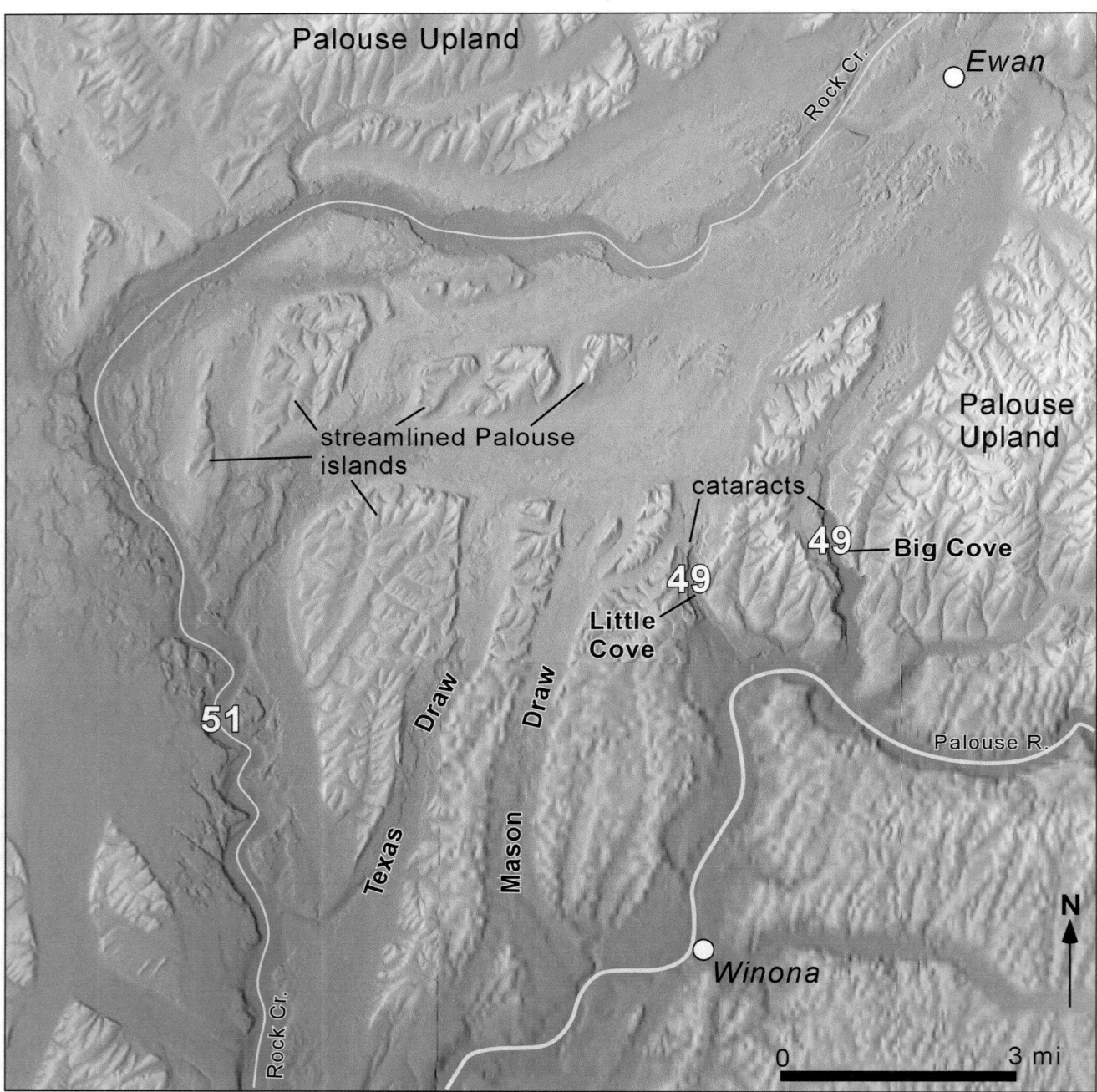

Palouse Upland

Ewan

Rock Cr.

Palouse
Upland

streamlined Palouse
islands

cataracts

49 — Big Cove

49

Little
Cove

51

Texas
Draw

Mason
Draw

Palouse R.

Rock Cr.

N

Winona

0   3 mi

Competing Coulees A series of semi-parallel flood coulees are located at the lower end of the Cheney-Palouse Tract. Big Cove Coulee along with Little Cove Coulee, Mason Draw and Texas Draw operated together to deliver the sudden onslaught of floodwaters to the Palouse River valley just downstream. These multiple channels were simultaneously occupied in order to discharge all the floodwaters that suddenly overwhelmed the drainage system. While some of the Palouse uplands were streamlined by the megafloods those at higher elevations were largely unaffected, like the uplands in white at upper right. Numbers correspond to features described on other pages.

Cove Coulees   Megafloods eroded deep recessional-cataract canyons at Little (above) and Big (below) Cove Coulees. Basalt bedrock was deeply incised between scarped Palouse hills on either side of the coulees. These parallel coulees simultaneously transferred floodwaters through a portion of the lower Cheney-Palouse Tract.

Rock Creek Coulee Above: Rock Lake, at the far east end of the Channeled Scabland, is its deepest lake. The flood-gouged rock basin is eight miles long and one mile wide. Submerged at the head of the lake lies a 400-ft deep plunge pool. Looking upstream to the north. Below: Rock Creek, only one of two perennial streams within the Cheney-Palouse Tract, today drains the eastern margin of the Channeled Scabland. Pale brown Palouse uplands in upper right generally lay above maximum flood levels. Since the last Missoula megaflood about 15,000 years ago only the extremely underfit Rock Creek trickles through the massive canyon after exiting Bonnie Lake (top center). Looking north into flood-flow direction.

Out-of-Place Meanders Along Lower Rock Creek Being one of the few perennial streams within the scabland, Rock Creek has managed to carve a few entrenched meanders into basalt bedrock, both between and since the time megafloods flowed through here. However, during megafloods the meandering canyon and everything else in this image would have been submerged under hundreds of feet of roaring floodwater. Like a firehose the deep floodwaters roared straight through – without much regard for a lazy, meandering river-cut canyon-eroding a wide swath of scabland on either side. Towel Falls of Rock Creek at lower right (arrow).

Lower Washtucna Coulee   Before the megafloods this coulee was initially cut by the Palouse River but later abandoned when megafloods carved a new, more-direct path to the Snake River via Palouse Falls (see page 56). Ever since, the lower Washtucna Coulee has been stream-less. Above: a bar of flood sediment formed in a more-protected part of the coulee in upper left of image. Below, bar is composed of mostly basaltic coarse sand and gravel. Like many other flood bars in the scabland an active borrow pit removes sand and gravel for use as aggregate. Notice the huge size of unwanted boulders, at center left, on the valley floor removed from the quarried flood bar. Also notice some faint giant current ripples along the top of the bar (arrow).

Washtucna Coulee Giant Current Ripples   Giant flood bars like this one, also along Washtucna Coulee, display some wonderful examples of giant current ripples composed primarily of flood-transported sand and gravel. Like most ripples they display an asymmetric profile with the steeper side of the ripple facing downstream. Top: arrow indicates flow direction for these ripples, which was down coulee to the right. Bottom: close up of borrow pit dug into the ripples shown above. Arrow points to the erosional contact between more-recent, gray-colored Missoula flood deposits that overlie pre-existing, tan-colored, Palouse loess underneath.

High-Energy Megaflood Deposits Another flood bar along Washtucna Coulee blanketed with giant current ripples (arrow indicates flow direction.) A close-up of the exposed interior of a borrow pit (outlined) is shown below. Chaotic stratification in the gray, basaltic sand and gravel indicates these sediments were deposited under extremely turbulent, high-energy flood conditions. On top is a thin cap of brown, wind-blown loess deposited since the last megaflood ~15,000 years ago.

Hijacked Palouse River During Ice Age floods the Palouse River was captured by the Snake River when mega-floodwaters spilled over a drainage divide along Washtucna Coulee (above). Since that time Upper Palouse Falls (below) marks the point of stream capture and dramatic shift in river direction. The new path of the Palouse River followed a long, deep tectonic fracture in the basalt preferentially hollowed out by the Missoula megafloods.

"Palouse River today is detoured from its former course along now streamless Washtucna Coulee into a striking, joint-determined scabland canyon which cuts through the pre-glacial divide." Bretz, Smith and Neff (1956)

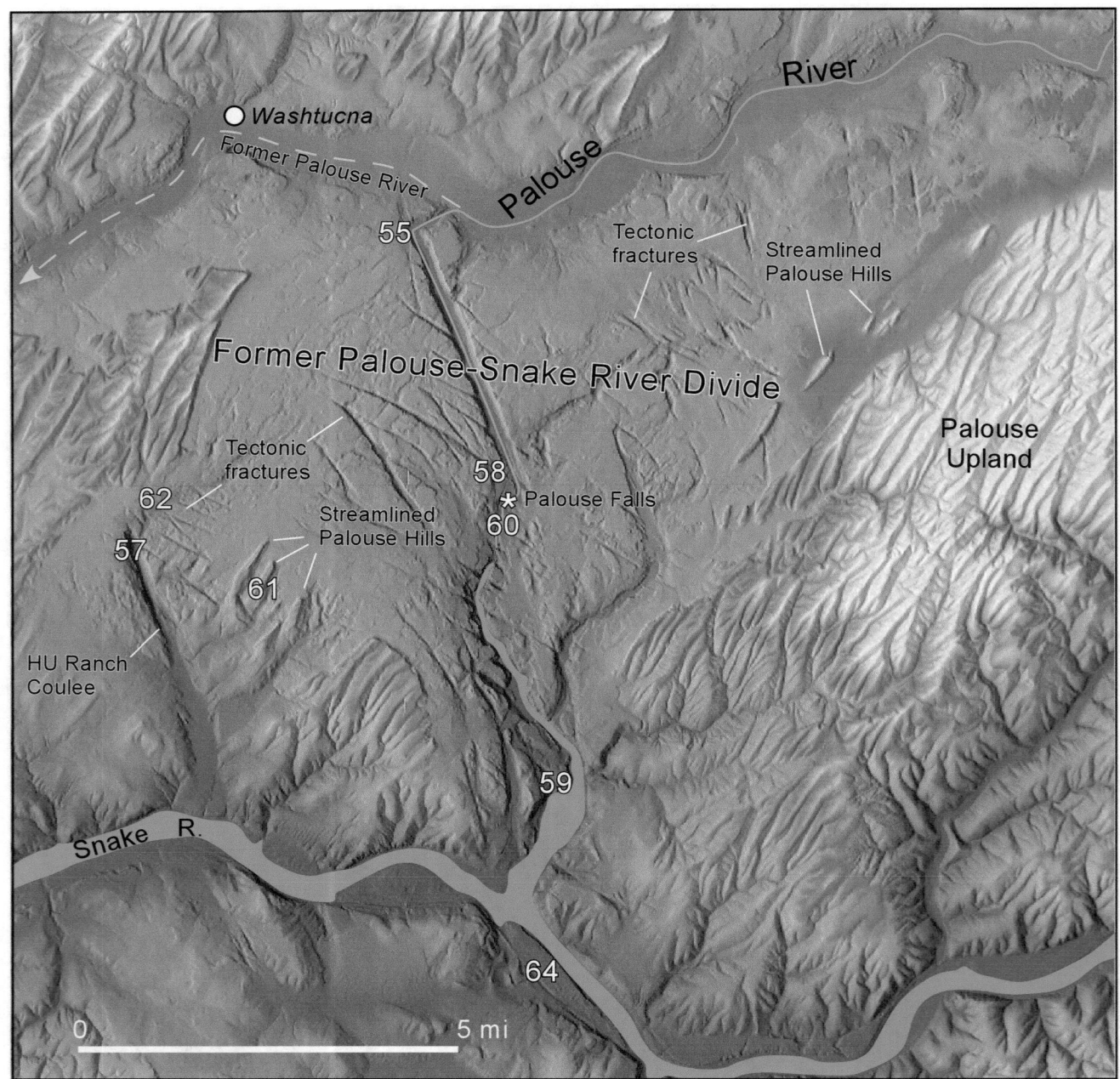

Ice-Age River of No Return   Powerful Missoula floodwaters, taking a short cut to the Snake River, carved a whole new canyon for the Palouse River. Since then, the river's course through the lower Washtucna Coulee was permanently cut off.

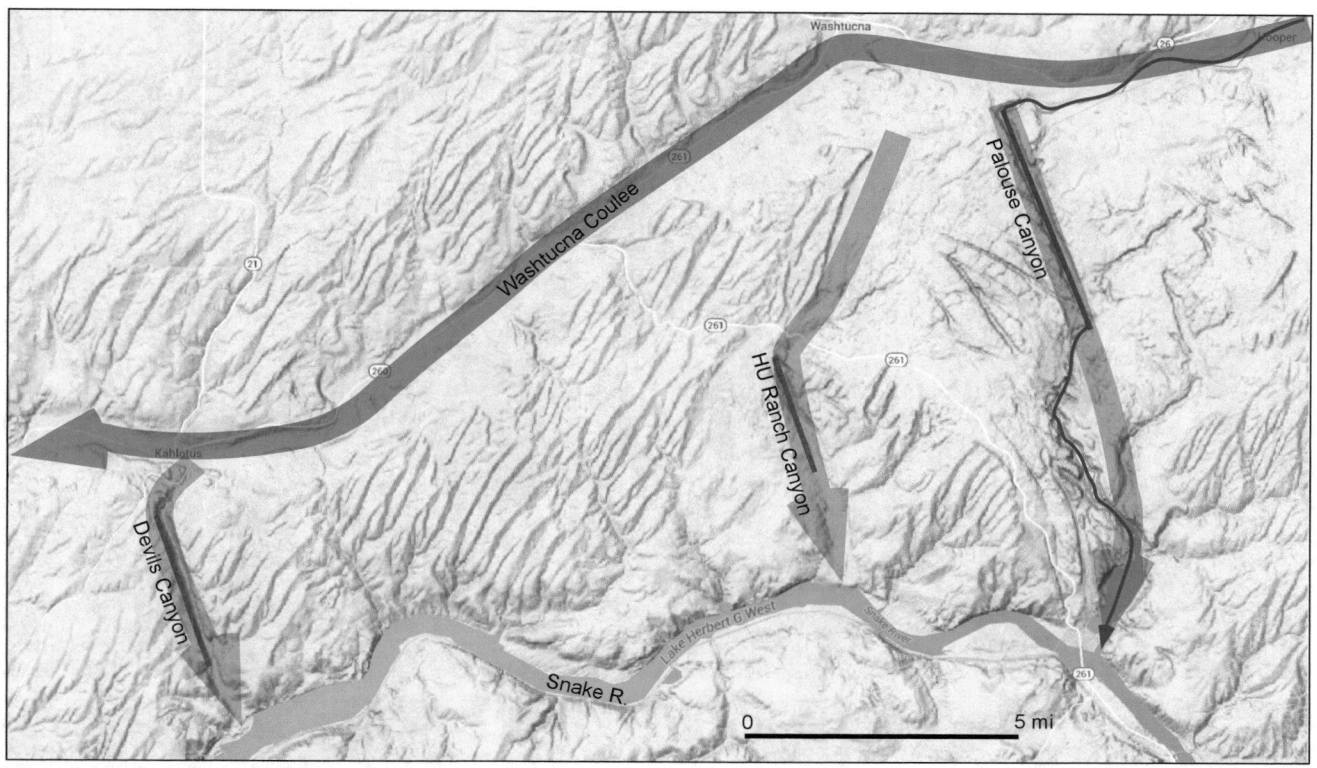

Unusually Straight Coulees   Above: the Palouse River, HU Ranch Coulee (below) and Devils Canyon are all deeply scoured flood coulees that are exactly parallel along a NNW trend. This is due to preferential erosion by floods along pre-existing rock joints or fractures in the basalt.

"Here is HU Ranch dry cataract, 280 feet from brink to bottom of plunge basin and that basin one and a half miles long in the bottom of the cataract's recessional gorge." Bretz (1959)

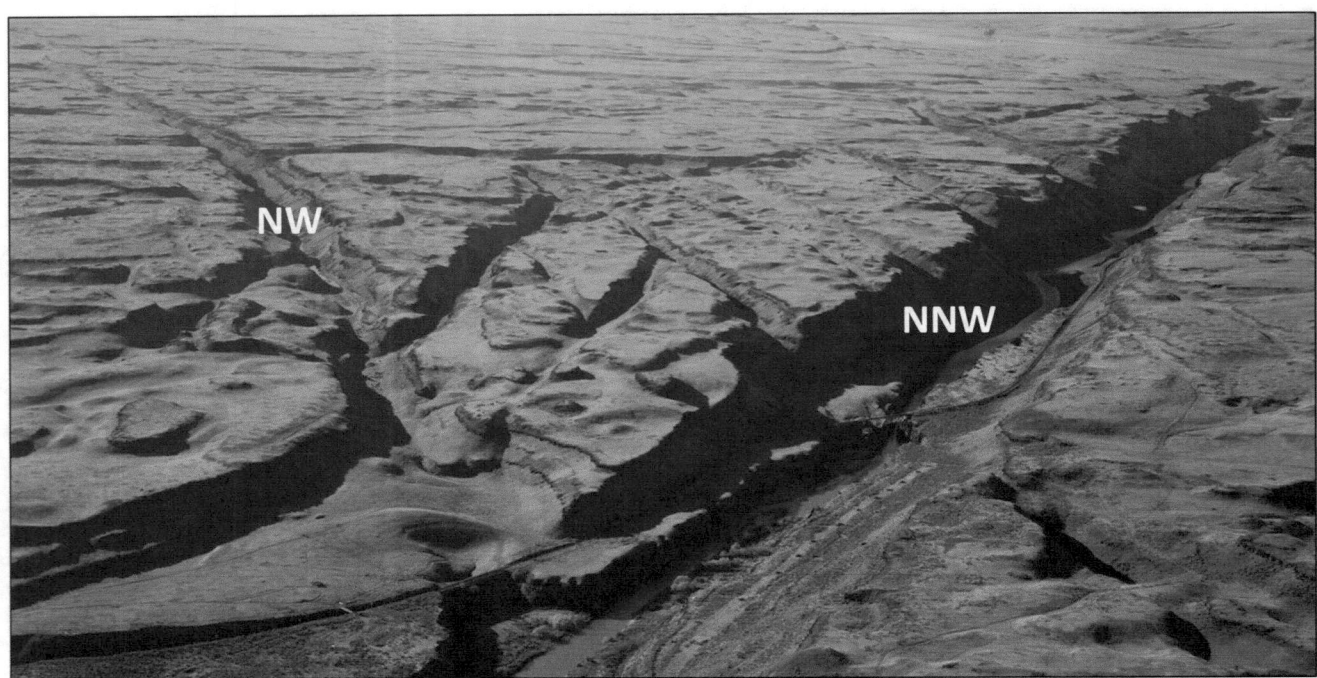

Scabland Plexus This area along the Palouse River canyon was once completely covered in a thick blanket of loess. Ice Age megafloods effectively stripped away all the Palouse soil before attacking the underlying basalt. Linear grooves developed where floodwaters preferentially eroded basalt that was weakened along two sets (NW and NNW) of deep tectonic fractures (joints) in the bedrock.

Palouse River Canyon After the Palouse River was captured by the Snake River a recessional cataract migrated upriver during multiple floods to create the canyon shown here. A wide range of erosional flood features is represented, including basalt mesas, buttes, and rock benches. Marmes Rockshelter (arrow) is, one of the oldest (~13,300 yrs) archeological sites in Washington. Today it, lies submerged inside a rocky berm unsuccessfully installed in 1969 to protect the site from rising waters associated with a hydroelectric dam downstream.

Washington State Waterfall Today's Palouse Falls drops ~200 feet over the lip of a cataract that, during megafloods, receded five miles upstream from the Snake River confluence. During megafloods everything in this image was under many hundreds of feet of floodwater.

Palouse Falls Then and Now Artist Stev Ominski's portrayal of Palouse Falls during the waning stages of a Missoula flood (left) versus today (right).

Palouse Armada Above: Like a flotilla of battleships, these streamlined Palouse hills stand guard near the entrance to Palouse Falls. Looking southeast toward the Snake River canyon. Block arrows indicate flood-flow direction. Below: two of the same loess hills viewed at ground level. The largest megafloods likely overtopped these aligned hills.

HU Ranch Coulee Here lies another deep coulee preferentially eroded into a straight and narrow canyon along deep tectonic fractures, exactly parallel to the Palouse and Devils Canyons (see page 57). During megafloods the canyon lengthened upstream by way of the recessional cataract visible just right of center. Today no active stream lies anywhere near the canyon. Another set of straight tectonic fractures, trending ~30-degrees away from the main coulee were also etched out during the same megafloods. Looking south toward the Snake River canyon, visible along top.

Devils Canyon Devils Canyon is another one of several parallel, arrow-straight canyons that delivered floodwater from Washtucna Coulee to the Snake River. To the left lie rolling, loess-covered Palouse hills planted in dry-land wheat fields—untouched by megafloods. Looking upstream into the direction of megaflood flow.

Down Devils Canyon   Above: At the town Kahlotus megafloods tearing down Washtucna Coulee flowed up and over this 100-ft divide (arrow) before they spilled over into stream-less Devils Canyon. Floodwaters that overfilled Washtucna Coulee spilled over into Devils Canyon along another weakened fracture zone. Like HU Ranch Canyon no water has flowed here since the last Missoula flood about 15,000 years ago. Below: the unwavering straightness of Devils Canyon.

Mid-Canyon Flood Bar In the foreground is a 2.5-mile-long, 200-ft tall flood bar that occupies more than half the width of the Snake River canyon floor. This huge bar lies across from the mouth of the Palouse River canyon. Here, floodwaters cascading down the Palouse River Canyon plowed into the high walls of the Snake River valley, forcing floodwaters to split before rushing in opposite directions. Some of the floodwater surged eastward up the Snake River for 80 miles all the way to Lewiston, Idaho. Most of the floodwaters, however, travelled west down the Snake River valley. Some weakly developed giant current ripples are visible atop the bar in the distance. The dimpled-looking surface closer up is patterned ground that has developed since the last Missoula flood ~15,000 years ago. Looking southeast.

Scott Flood Bar Undulating giant current ripples along the Snake River, looking downstream (southwest). The asymmetric ripple profile is apparent in shadow along the railroad cut in foreground.

Lake Sacajawea Flood Bar An exposure, up to 400-ft tall, of coarse-textured flood deposits lies along the Snake River where flood sediment gathered and was preserved as a pendant-eddy-type bar just downstream of a basalt spur. Below: floodwaters rose hundreds of feet above the bar as indicated by eroded escarpments along Palouse hills. Looking upstream. See also page 19.

"A great bar along the Spokane, Portland and Seattle Railroad, 2.5 miles northeast of Snake River Junction, is 400 feet thick. The gravel pit excavated by the railroad here has a scarp 200 feet high. The gravel is unindurated and slides easily, so that the structure is difficult to decipher. But it is 99 per cent basalt." Bretz (1925)

Telford-Crab Creek Tract

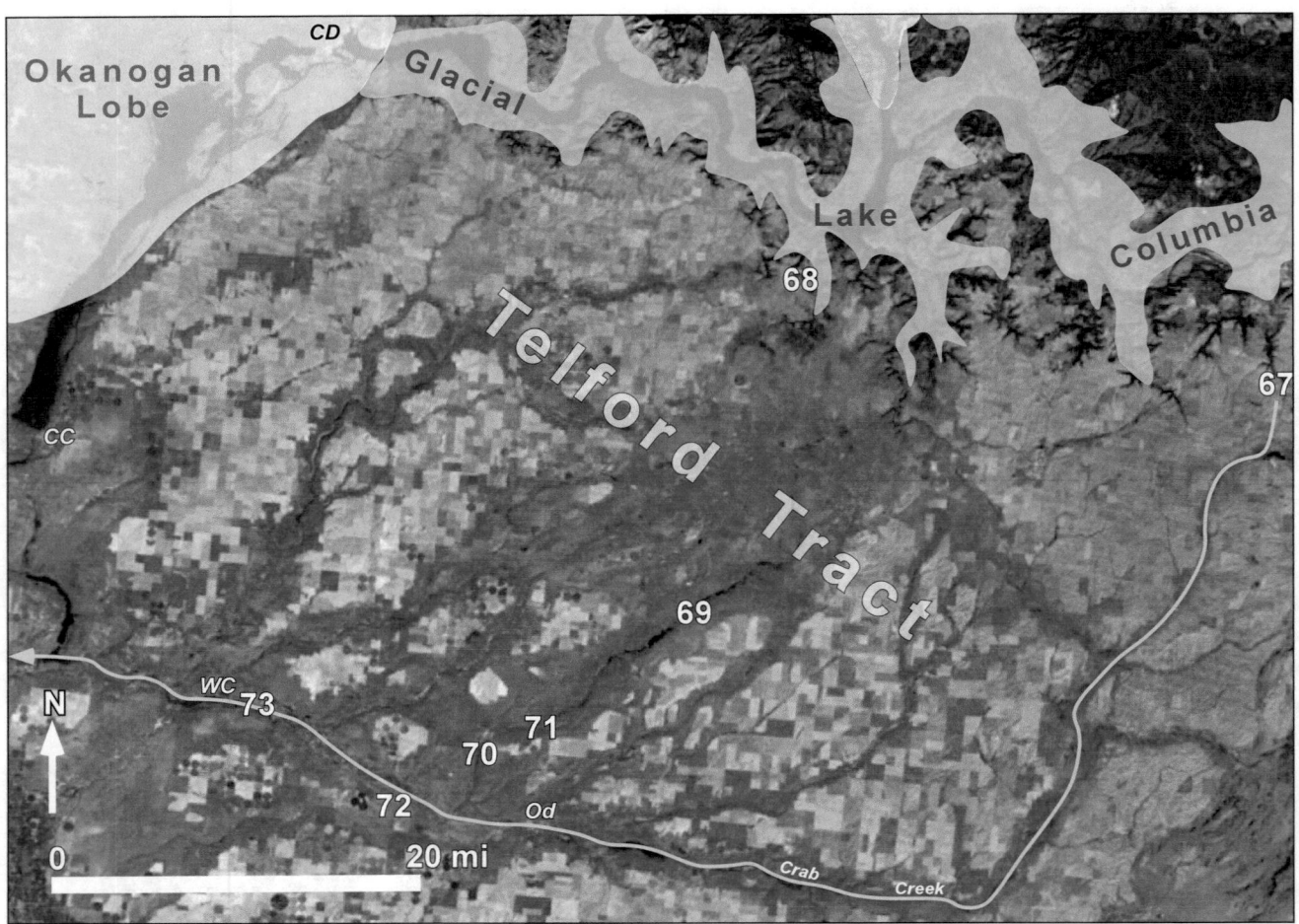

Index Map for the Telford-Crab Creek Scabland Tract The Telford-Crab Creek Tract is a collection of anastomosing dark flood channels in basalt located between the Cheney-Palouse Tract (lower right) and Grand Coulee (upper left). Blockage of the Columbia River by the Okanogan Lobe and subsequent formation of glacial Lake Columbia caused floodwaters to spill over low divides (~2300 ft elev) at the head of the Telford Tract. Numbers correspond to flood features on pages described herein. Od = Odessa, WC = Wilson Creek, CC = Coulee City, CD=Coulee Dam.

"That all the channels of this Telford-Crab Creek spillway were in operation at one time cannot be gainsaid." Bretz (1928a)

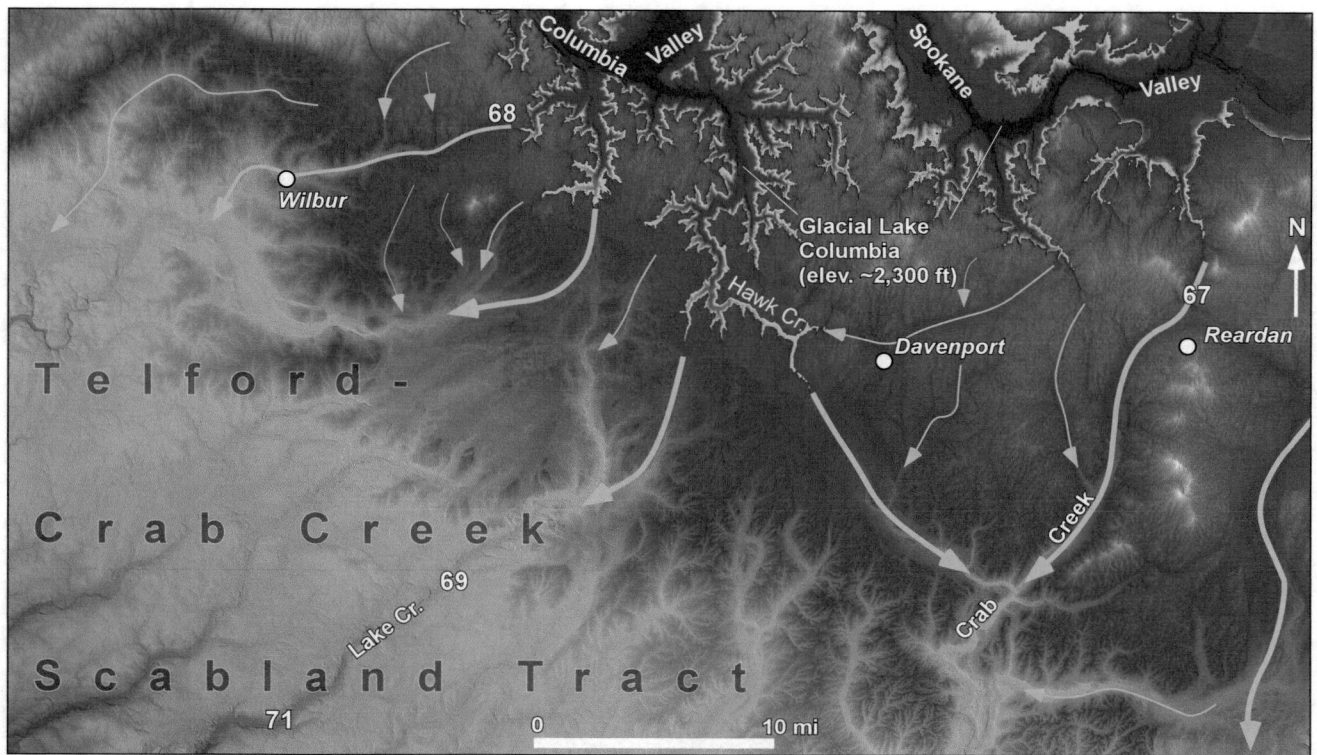

Missoula Floods Spillover from Glacial Lake Columbia At times, during the Ice Age, glacial Lake Columbia filled the Columbia and Spokane River valleys up to an elevation of 2300 ft. When megafloods from glacial Lake Missoula invaded from the east they ran over Lake Columbia rising up to ~2500 ft elevation – instantly spilling across low divides at the head of the Telford-Crab Creek Tract (arrows). Highest elevations on this map are white, lowest green.

Flood Spillover at Head of Crab Creek A shallow spillover channel (arrow) cuts through farmlands at the head of Crab Creek near Reardan as Missoula megafloods overfilled glacial Lake Columbia. Only a thin mantle of windblown soil covers basalt bedrock on either side of the spillover channel. Looking north towards former Lake Columbia.

Ice-Rafted Erratics at Head of Telford Tract Lake Roosevelt (reservoir behind Grand Coulee Dam) partially fills the Columbia valley in the distance. During most of the last glacial cycle this same valley was formerly occupied by much-larger glacial Lake Columbia (up to 1000 ft higher than Lake Roosevelt). Lake Columbia resided here for thousands of years when the Okanogan Lobe blocked the Columbia River to the west. Missoula megafloods that invaded Lake Columbia during this time imported exotic boulders (erratics) floating in icebergs. Both erratics shown here rest on basalt bedrock. Above is a granitic boulder grounded at 2475 ft elevation about 200 ft above the maximum level of former glacial Lake Columbia. Below is a boulder of banded argillite traceable back to the ice dam for glacial Lake Missoula over 100 miles away.

Upper Lake Creek Coulee Many natural, flood-scoured water bodies lie along Lake Creek Coulee including Coffeepot Lake (above) and Twin Lakes with a giant, circular pothole (below).

Lower Lake Creek Coulee Flood coulees naturally grow deeper and wider downstream where they are often lined with multiple, stair-step-like rock benches.

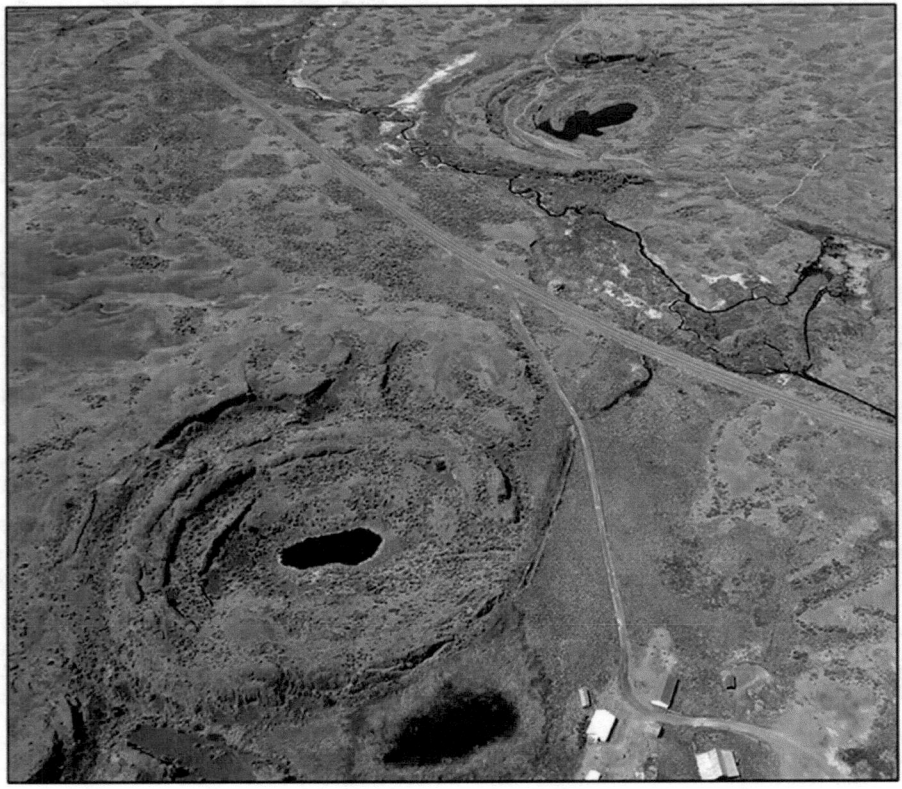

Odessa Ringed Craters along Lower Lake Creek Coulee Unusual rings in the basalt bedrock formed as a 15-million-year-old lava flow cooled. Much later the basalt rings were covered over by windlblown Palouse loess. Subsequent erosion by megafloods exhumed the buried basalt surface. Today, ringed craters stand out due to differential erosion by floodwaters of craters and discordant dikes within the basalt bedrock. The greatest concentration of ringed craters exists within the Telford-Crab Creek Tract here near Odessa.

Amphitheater Crater This crater consists of multiple, concentric, ring dikes of basalt.

Cache Crater This unusual crater, hollowed out by Ice Age floods, is surrounded by only a single raised berm of basalt.

Series of Giant Current Ripples Along Crab Creek   Perhaps the highest concentration of giant ripples exists along Crab Creek between Odessa and Wilson Creek. Flow direction (arrow) is indicated by asymmetry, with steeper sides of ripples pointing downstream. Ripples are composed of primarily loose, basaltic, coarse gravel and sand transported and deposited by the megafloods.

Multiple Sets of Giant Current Ripples   Deep, fast-moving floodwaters simultaneously deposited ripples onto at least three different levels here along Crab Creek Coulee.

"Giant current ripples on gravel bars, difficult to identify at ground level under a cover of sage brush." Bretz (1969)

Grand Coulee

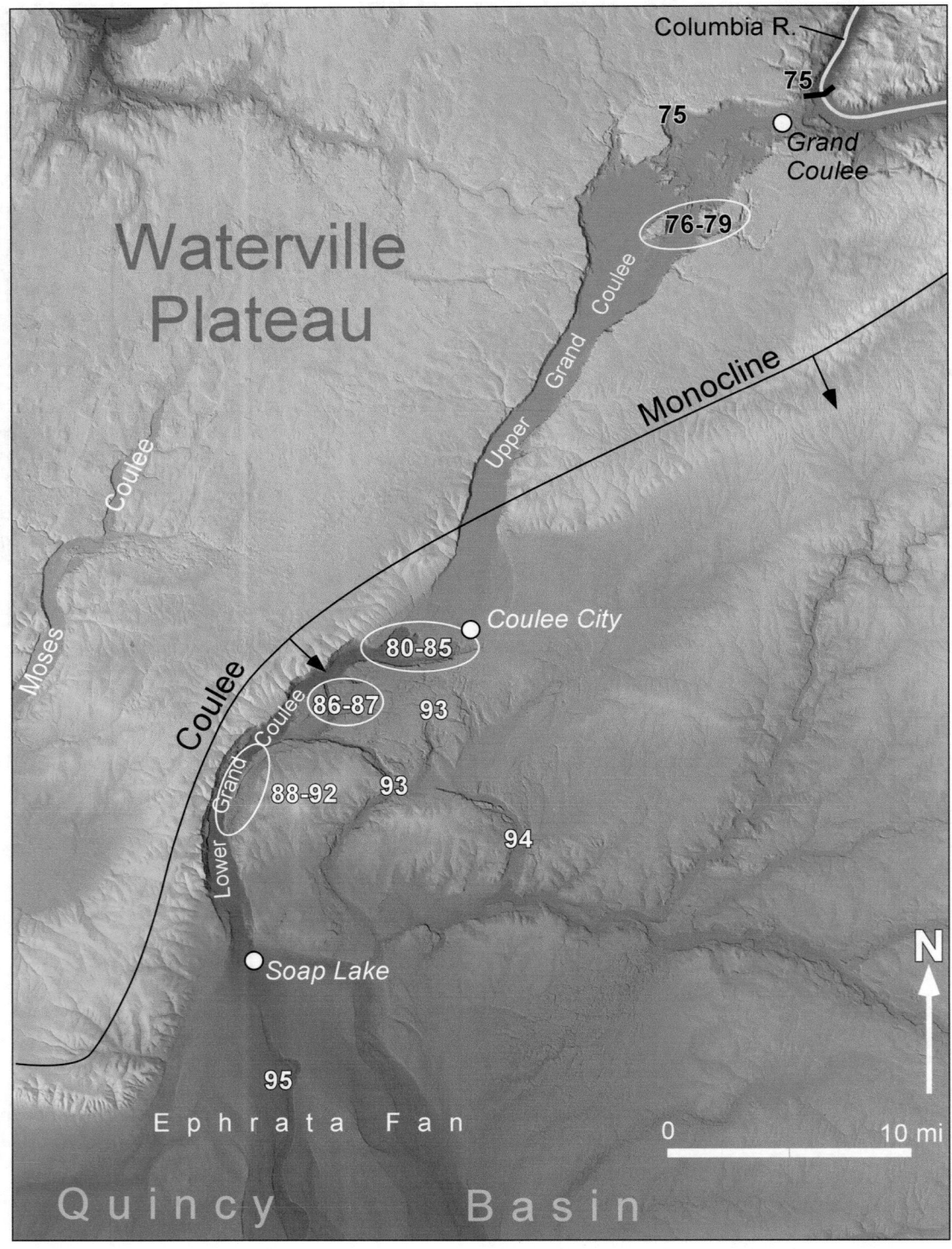

Grand Coulee Index Map Grand Coulee starts near the Grand Coulee Dam (next page) and stretches for 50 miles to Soap Lake where the coulee suddenly expands into the Quincy Basin. During most of the last glacial cycle the Columbia River was blocked by the Okanogan Ice Lobe that advanced onto the Waterville Plateau, forcing floodwaters from glacial Lake Missoula down Grand Coulee. Initially, the spillway for megafloods down Grand Coulee was much higher but lowered suddenly by 900 ft when a recessional cataract cut through the head of the coulee. In between Missoula floods the Columbia River was also diverted down Grand Coulee by the Okanogan Lobe, which held in place for most of the Missoula floods. Numbers correspond to flood features on pages described herein.

Head of Grand Coulee The upper end of Grand Coulee hangs 500 feet above the Columbia River at lower left. Irrigation water from Lake Roosevelt, backed up behind Grand Coulee Dam, is pumped up into the coulee to fill Banks Lake at upper right. Steamboat Rock (arrow) is a remnant of the basalt bedrock that once filled Grand Coulee prior to the Ice Age floods.

Upper Grand Coulee The flat basalt surfaces on either side of Grand Coulee, as well as Steamboat Rock were all once connected. Grand Coulee began to form when the Okanogan Lobe advanced onto the Waterville Plateau, blocking the Columbia River. Subsequent Missoula floods were forced to flow along the margin of the ice sheet. A cataract that began 20 miles downvalley receded upstream during each Missoula flood. Eventually the cataract self destructed upon breaking through the head of the coulee, leaving behind the 900-ft deep canyon we see today.

Emerald Green Banks Lake   Views across upper Grand Coulee from the north (above) and south (below).

"When the cascade migrated north... it became a typical recessional cataract (Steamboat Falls) nearly 900 feet high. Eventually its retreat extended the lengthening gorge across the divide into the preglacial Columbia River Valley, and thus the cataract destroyed itself." Bretz, Smith and Neff (1956)

Banks Lake and Steamboat Rock   Man-made dams exist at both the upper and lower ends of the upper Grand Coulee to contain irrigation water in Banks Lake. This water, pumped up from Lake Roosevelt, is distributed over a large part of the mid-Columbia Basin to support the expansive agricultural industry. There is good geologic evidence that the Okanogan Ice Lobe filled the upper Grand Coulee toward the end of the Ice Age. This is demonstrated by both: (1) glacial erratics overlying basalt atop Steamboat Rock, including huge granitic boulders like the one above and (2) glacially polished and striated (arrows) surface of granitic bedrock at lake level near Steamboat Rock (next page).

Castle Rock Cataracts and Northrup Canyon   A pair of short recessional cataract canyons (arrows), near the mouth of Northrup Canyon, were created by the spillover of floodwater that once filled upper Grand Coulee in the background. Granite bedrock crops out below the basalt at lower elevations within the canyon. Looking north toward head of Grand Coulee (upper left). Next page: Megafloods also spilled over into the upper Grand Coulee from the northeast via Northrup Canyon. Secluded heart-shaped Northrup Lake lies at the head of one arm of Northrup Canyon.

Great Cataract Group   About halfway down Grand Coulee is a group of recessional cataracts, altogether four miles wide. Iconic Dry Falls (below) is located at the western part of this cataract complex. The Great Cataract Group divides the upper from lower Grand Coulee where flood-waters dropped 400 ft over the lip of Dry Falls into several plunge-pool basins below. Numbers correspond to flood features described on pages herein.

Dry Falls

Dry Falls Then and Now Ice Age floods descending Grand Coulee dropped over a 40-story-high waterfall that altogether receded 25 miles from the mouth of the coulee near Soap Lake. During its peak, water flowing over the rock precipice was another 40-stories deep. Floodwaters stripped bare the basalt surface before scarring with giant grooves and potholes. During the Ice Age the Okanogan Ice Lobe on the Waterville Plateau (left) came to within a few miles of Dry Falls based on the still-visible remains of the Withrow Moraine. Multiple alcoves formed from as many flood-carved cataracts that receded simultaneously. The Umatilla Rock blade, an erosional remnant of basalt, extends below two of the horse-shoe shaped alcoves. Artwork by Stev Ominski.

Dry Falls Cataract Complex Three alcoves simultaneously transported floodwaters over the lip of the falls as it excavated its way upstream. Giant grooves and circular potholes scar the surface behind the 400-ft tall cataract cliff. See also page 14.

Umatilla Rock Blade As massive floodwaters spilled over Dry Falls their erosive power split into multiple conduits across the broad lip of the falls, ultimately preserving this rock blade down the middle.

Castle Lake Cataract and Plunge Pool   At the east end of the Great Cataract Group (see page 80), two miles east of Dry Falls is the Castle Lake Cataract. The almost-dry Castle Lake, a plunge-pool depression, lies at the base of the 300-ft tall cataract. The Castle Lake basin drops into Deep Lake Coulee at upper right. All irrigation water for the Columbia Basin Project, coming out of Banks Lake, runs underground for two miles via two, giant, underground tunnels below the broad, flood-swept plateau at the top of this image.

"Castle Lake Falls totals about 300 feet in height above Deep Lake ... the only human uses of this desolate area south of Coulee City are traverses, gravel pits, springs, the main irrigation canal, and perhaps jackrabbit hunting." Bretz (1969)

Deep Lake Coulee Deep Lake Coulee empties into the lower Grand Coulee just below Dry Falls and Umatilla Rock. This photo looks west toward the entrance into lower Grand Coulee. Notice giant grooves (arrows) scored into the basalt bedrock on uplands along either side of Deep Lake Coulee. These grooves trend cross wise with the coulee suggesting they predate the formation of Deep Lake Coulee when the upper basalt surface continued, uninterrupted, across the coulee.

Megapotholes at Deep Lake Some of the best examples of giant, cavernous potholes exist near Deep Lake. Giant potholes appear to have been drilled into basalt during the megafloods via giant circulating eddies created by flood turbulence called "kolks" – a type of subfluvial tornado.

"Deep Lake, below one of the Grand Coulee abandoned falls, has many associated huge potholes, drilled into the basalt at the foot of the falls as they retreated." Bretz (1923)

Closeup of Giant Pothole in East Lenore Coulee.

Flood-Tortured Basalt Below Dry Falls   Severely eroded and chaotic butte and basin scabland spreads out across a 30-square-mile area below Dry Falls – visible at upper left.

"Canyons in the scablands are multiple and anastomosing, amazingly so in some tracts; deep canyons and shallow ones uniting and dividing in a labyrinthine fashion about bare rock knobs and buttes unlike any other land surfaces on the earth." Bretz (1927)

Dualing Recessional Cataract Canyons Two adjacent recessional cataracts battled for real estate as they eroded upstream from the edge of Jasper Canyon. During megafloods this entire landscape was submerged beneath hundreds of feet of turbulent floodwater. Besides the recessional cataracts, note chaotic distribution of giant potholes, grooves, and rock benches. Multiple rock benches were created as megafloods peeled back edges of once-continuous basalt flows like the layers of an onion. Looking northeast.

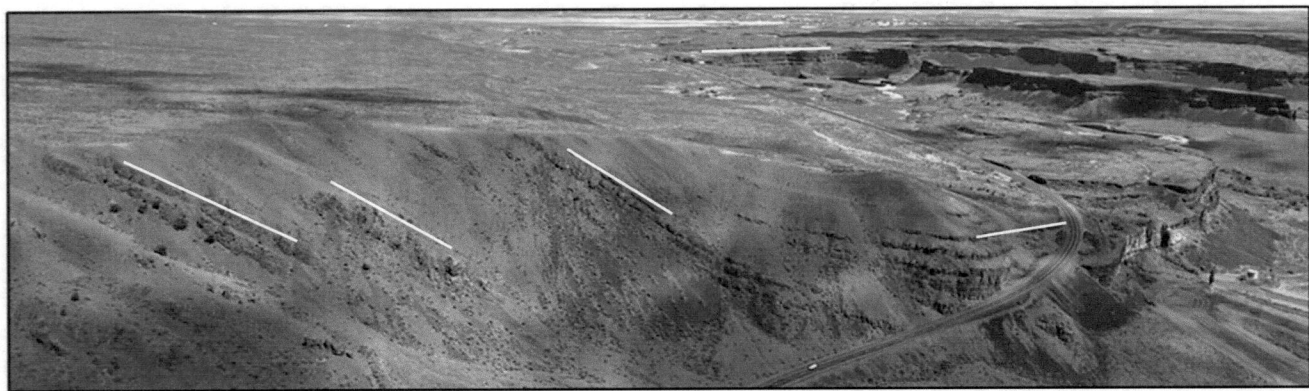

Coulee Monocline   The Coulee Monocline is a folded geologic structure that runs along the west side of lower Grand Coulee (see page 74). The monocline is visible where steeply dipping layers of basalt (originally horizontal) have been tectonically pushed upward (tilted white lines above). Upper photo looks north toward Dry Falls where lava flows are essentially horizontal at the base of the monocline. Tectonic uplift along the Coulee Monocline created hogback islands (arrows below) scoured by the floods within today's Alkali Lake. Here, the originally flat basalt flows have been tilted ~45 degrees. Tectonic movements along the monocline all occurred prior to the Ice Age floods. In the shadows (upper left) are castle-like silhouettes of the Great Blade.

Lakes of Lower Grand Coulee A string of beautiful lakes occupies almost the entire length of the lower coulee. Both above and below is dramatic erosional scabland on either side of Lake Lenore, looking south. Both images show extreme and chaotic erosion on rock benches hundreds of feet above Lake Lenore. A series of hanging valleys visible along the top of the lower image were left behind after megafloods preferentially eroded the more fractured and weakened rock along the Coulee Monocline.

Alluring Lake Lenore Islands within Lake Lenore are hogbacks composed of basalt that dip steeply to the right – again part of the folded Coulee Monocline. Because of the displacement along the monocline the same basalt flows on top of the plateau on the left are several hundred feet lower on the right side of the lake.

The Great Blade   The Great Blade runs for almost four miles along the east side of lower Grand Coulee. It is the longest blade of basalt anywhere within the Channeled Scabland. Similar to Umatilla Rock at Dry Falls the Great Blade developed as an erosional remnant of two parallel cataract canyons that receded simultaneously upstream. Many other erosional flood features including potholes, grooves, and flood-swept mesas lie along both sides of the blade, including East Lenore Coulee on the right. Maximum relief on the 4-mile-long blade is 560 ft and in places the blade narrows to only a dozen feet wide.

"There were a few double falls each member of which receded at approximately the same rate, so that the island in mid-channel became very much elongated, like a great blade, as the falls receded and the canyons lengthened."
Bretz (1928a)

Out of Place   This light-colored granitic boulder stands out in stark contrast to the surrounding dark basalt bedrock. Ice-rafted erratics like this one are rare along high-energy flood environments like those along the Great Blade.

Alligator Point   Seen here is massively potholed rock bench that extends below the Great Blade – visible at top center. Looking northwest with Lake Lenore on the left, East Lenore Coulee on the right. Arrow points to perhaps the largest pothole (800 ft wide and 60 ft deep) observed anywhere in the scabland.

Hudson Coulee Recessional Cataracts Hudson Coulee is another example where two, horse-shoe-shaped, recessional cataracts eroded leaving behind a narrow rock blade (arrow) down the middle. Looking north towards Banks Lake and the upper Grand Coulee.

Dry Coulee After exiting the upper Grand Coulee some of the floodwaters spread out to the east via Dry Coulee. In this scene Dry Coulee wraps 180-degrees around the north, east and south sides of dome-shaped High Hill. During megafloods High Hill was an island that protruded above the highest flood levels and thus its summit was protected from erosional scouring. Therefore, only the upper slopes of High Hill support agriculture due to the thin mantle of windblown loess preserved there. Looking west.

Spring Coulee (aka Billy Clapp Lake)   Above: Spring Coulee is another side channel that splayed eastward from lower Grand Coulee. Today Spring Coulee holds irrigation water within the Billy Clapp Lake reservoir. While some of the invading floodwaters entered from the upper left, other anastomosing flood channels entered from the upper right (arrows). Below: undulating, giant current ripples lie on a sediment-covered flood bar above the reservoir. The extremely coarse-grained basaltic flood deposits on the bar do not retain much moisture or nutrients for farming – suggested by the apparent poor health of the crop.

Ephrata Fan The Grand Coulee ends at Soap Lake where floodwaters suddenly expanded into the broad Quincy Basin causing a sudden drop in the velocity of floodwaters. Today many huge boulders, some the size of small houses, lie scattered across the fan. Most are composed of basalt ripped out of Grand Coulee but there is also a small percentage of granitic boulders eroded from the upper end of the coulee. Rounding of boulders could be due to abrasion via rolling and bouncing along the base of the flood flow and/or inherited from in-situ surficial weathering at their upstream place of origin. Below: An especially large boulder of flood-deposited basalt entablature (circled above) is just one of thousands that came to rest on the fan.

"Grand Coulee water, on emerging from its rock-bound course, spread widely in Quincy basin, and the extensive gravel deposit there records a great decrease in transporting ability." Bretz, Smith, and Neff (1956)

Moses Coulee

Moses Coulee A fourth major pathway for Ice Age megafloods across the Channeled Scabland was Moses Coulee. Next to Grand Coulee, 50-mile-long Moses Coulee, with its 800-foot-tall walls, is the longest and deepest of the scabland coulees. Lower Moses Coulee shown here appears to have started as a "normal" stream valley. This is indicated by a long series of hanging valleys (in shadow) perched above the reamed-out and box-shaped coulee. Hanging valleys are the result of a sudden, rapid widening and perhaps deepening of the coulee. These hanging valleys formed in a similar way as those along the lower Grand Coulee – from lazy tributary streams that drained off the higher ridges prior to the megafloods. Outburst flood(s) quickly over-deepened and -widened the coulee, which left the side valleys hanging up to 600 feet above the floor of the coulee. Altogether up to five separate Ice Age flood events down Moses Coulee are inferred based on an analysis of flood deposits at the mouth of this coulee. Little or no water exists here in the lower coulee except for a tiny ephemeral stream.

"The empty canyons and dry cataracts of the plateau are easily the most conspicuous scabland forms. Among the canyons Grand Coulee takes precedence, followed by Moses Coulee." Bretz (1928a)

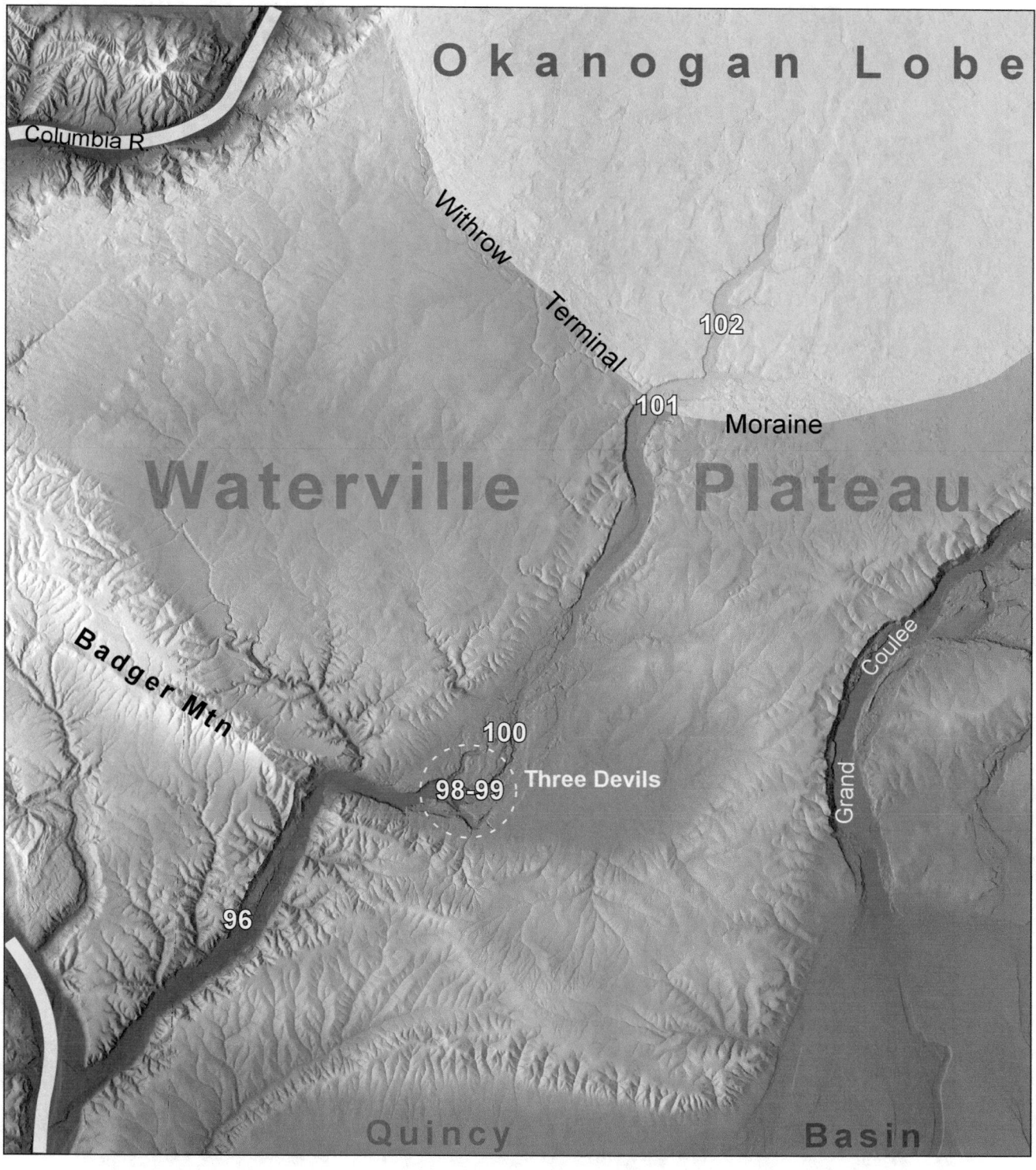

Three Devils Scabland Complex Midway along Moses Coulee is a maze of channels and coulees known as the Three Devils. Three Devils empty into a sole, flat-bottomed lower Moses Coulee, 800 ft deep, all the way to the mouth of the coulee. Numbers correspond to page numbers herein.

Coulee Transition at Three Devils Looking downstream across the southern arm of the Three Devils complex. Lower Moses Coulee neatly cuts across the Badger Mountain uplift near top center where it transitions to a single channel. Arrow points to head of recessional cataract. Unpaved portion of the solitary road (public) passes through most of the coulee's length.

Broad Scabland Complex Above Three Devils   Looking upstream (northeast), a maze of channels spreads out across the upper end of the Three Devils. The coulee complex returns to a mostly single flood channel at top right corner. See also page 15.

"Only extraordinary flooding could have crossed the violated preglacial divides, and only extraordinary velocity (born of huge volume) could scarify the bedrocks so tremendously." Bretz (1969)

Pristine Withrow Moraine Crosses Moses Coulee   At lower left lies undisturbed hummocky debris of the Withrow Moraine as it piled up along the front of the Okanogan Lobe glacier thousands of years ago. Down valley, immediately in front of and to the right of the moraine, are the braided remnants of curving meltwater stream channels that drained down coulee away from the front of the glacier called glacial outwash. Along the valley to the left is a megaflood bar, partially buried beneath younger glacial deposits. This is indisputable evidence that Moses Coulee is an older flood coulee that developed BEFORE the Okanogan lobe reached its maximum around 15–16 thousand years ago. Looking southwest down upper Moses Coulee.

"The Withrow moraine crossing of Moses Coulee is unique in all scabland. It is a very strongly expressed ridge or series of ridge hills of glacial drift marking the extreme southern limit ever reached by the northern ice sheet on this part of the plateau." Bretz (1959)

Upper Moses Coulee North of the Withrow Moraine sits Jameson Lake along the upper, shallower end of the Moses Coulee. This upper end is covered in glacial till left behind by the Okanogan Lobe AFTER the coulee was already carved out. In the distance, above Jameson Lake, the upper end of Moses Coulee loses definition and is hardly discernable upon the generally flat Waterville Plateau. Looking northeast.

Moses Coulee Conundrum Unlike most other coulees of the Channeled Scabland there is presently no obvious source of floodwater for Moses Coulee. The head of the coulee merges with the Waterville Plateau without any apparent connection with megafloods from Glacial Missoula coming from east. The lack of any soil development on the moraine or flood deposits within the coulee suggests that Moses Coulee formed from outburst floods that occurred earlier in the last glacial cycle, as the Okanogan Lobe was advancing across the Waterville Plateau, but not during any of the previous glacial cycles. A common interpretation has been that the coulee was cut by a Missoula flood or floods that flowed along the eastern margin of the Okanogan Lobe as it was advancing – before the glacial maximum >15 thousand years ago. This interpretation would also require that a deepened Grand Coulee was not present just to east – otherwise all Missoula floodwaters in the area would have drained that way instead. Complicating matters is the lack of any clear connection on the Waterville Plateau at the head of Moses Coulee with the Missoula floods coming from the northeast (e.g., incised channels across the plateau leading to Moses Coulee). Evidence here in Moses Coulee (e.g., Withrow Moraine) therefore suggest that floods occurred not only as the glaciers retreated, but also as the glaciers advanced, so that floods could have occurred at any time during a glacial cycle – not just during waning stages. Another more recently proposed source is a flood or floods that suddenly escaped from beneath the Okanogan Lobe glacier itself.

Quincy-Othello Basins

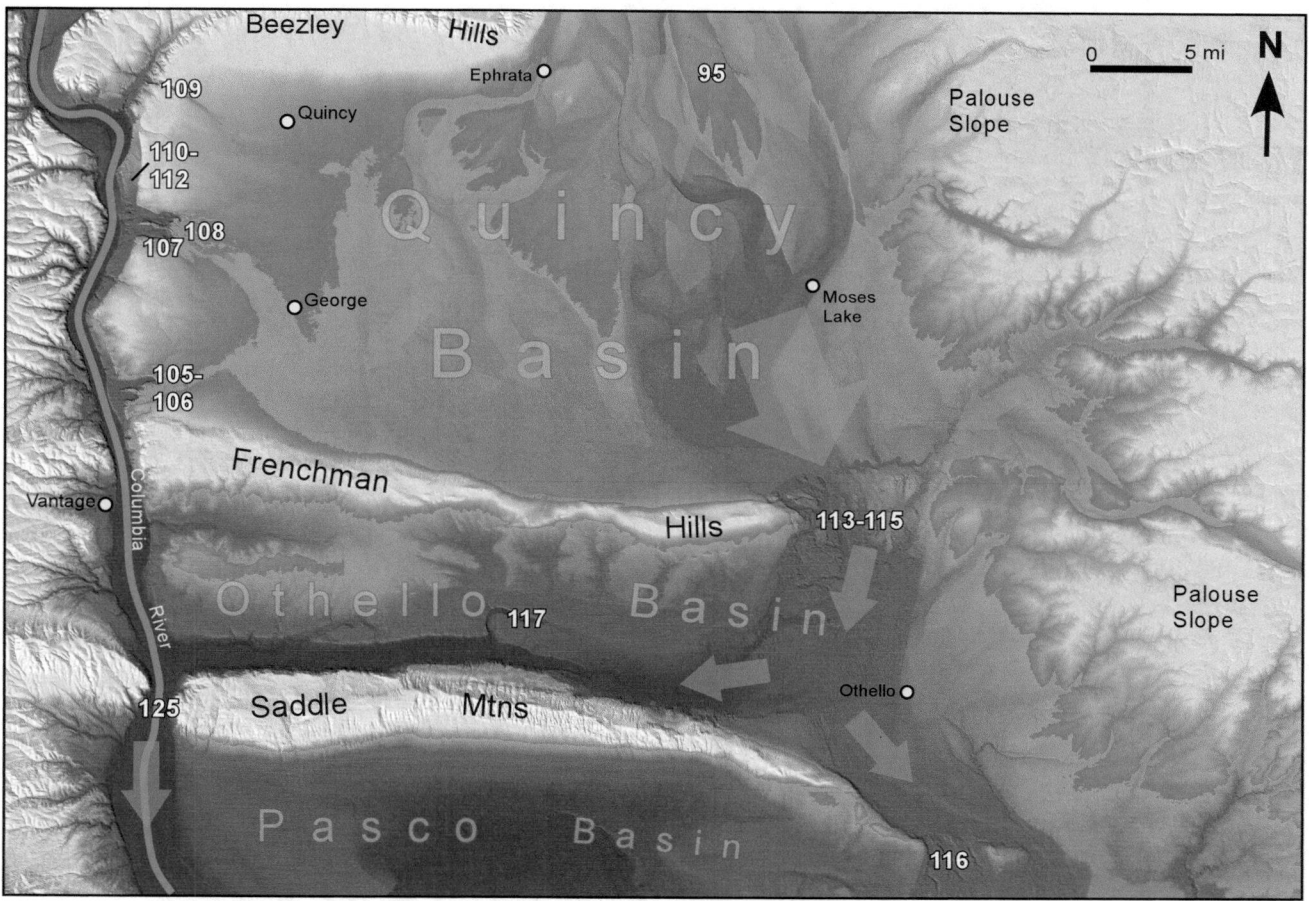

Index Map for the Quincy and Othello Basins Missoula floods invading the Quincy Basin entered from the north via Grand Coulee and east by way of Crab Creek. In southern portion of this map, white areas were generally above the maximum flood level. Numbers correspond to page numbers herein.

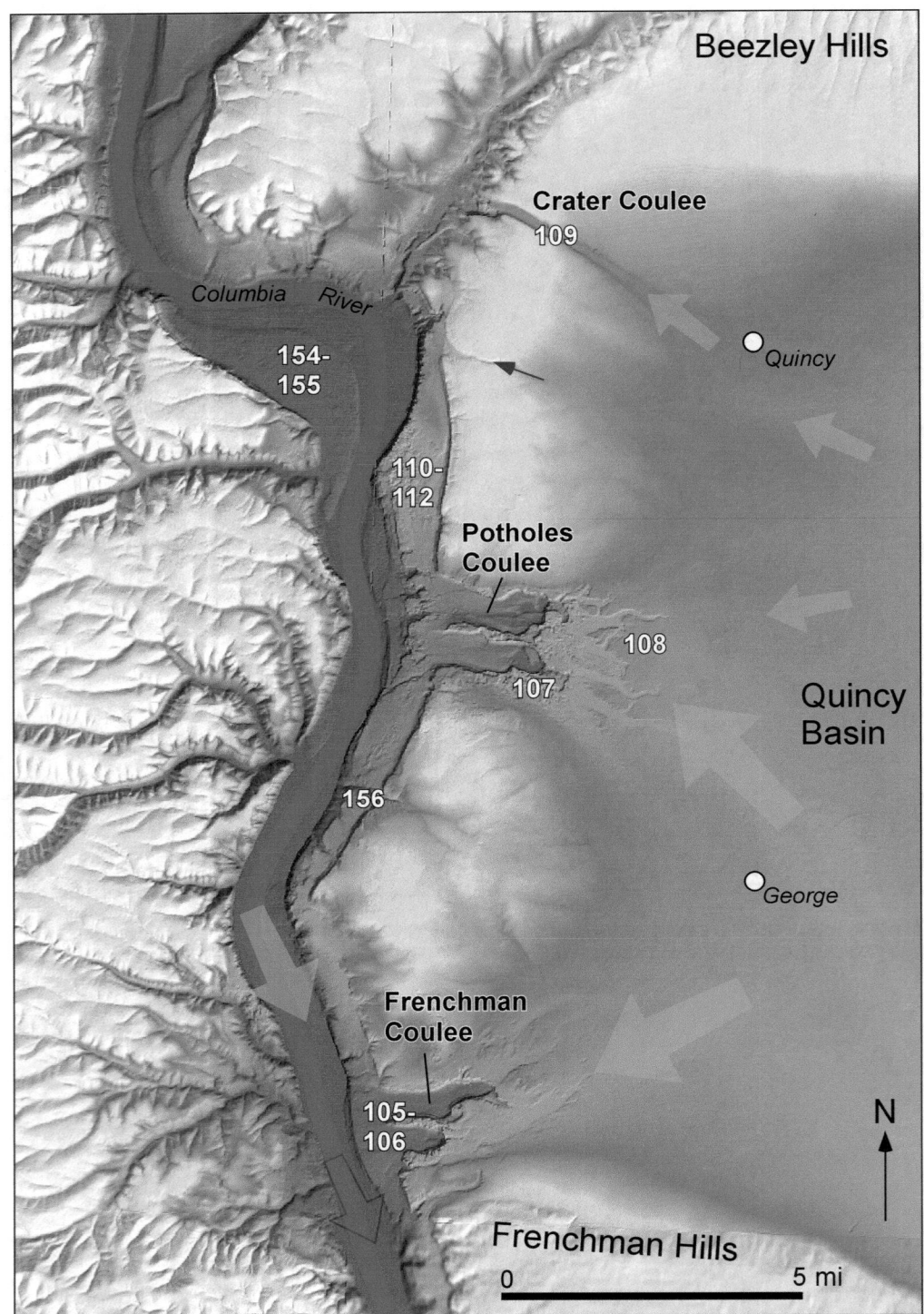

Quincy Basin Cataracts Missoula floods quickly overfilled the Quincy Basin causing the excess floodwater to spill over the west side at Frenchman, Potholes and Crater Coulees before cascading into the Columbia River hundreds of feet below. White areas were above maximum flood level. Blue arrows indicate Missoula flood-flow direction.

"Two other outlets for the Quincy basin existed during the Spokane epoch. They are at Frenchman Springs and the 'The Potholes,' two great notches in the wall of the Columbia Valley on the western margin of the plateau." Bretz (1923)

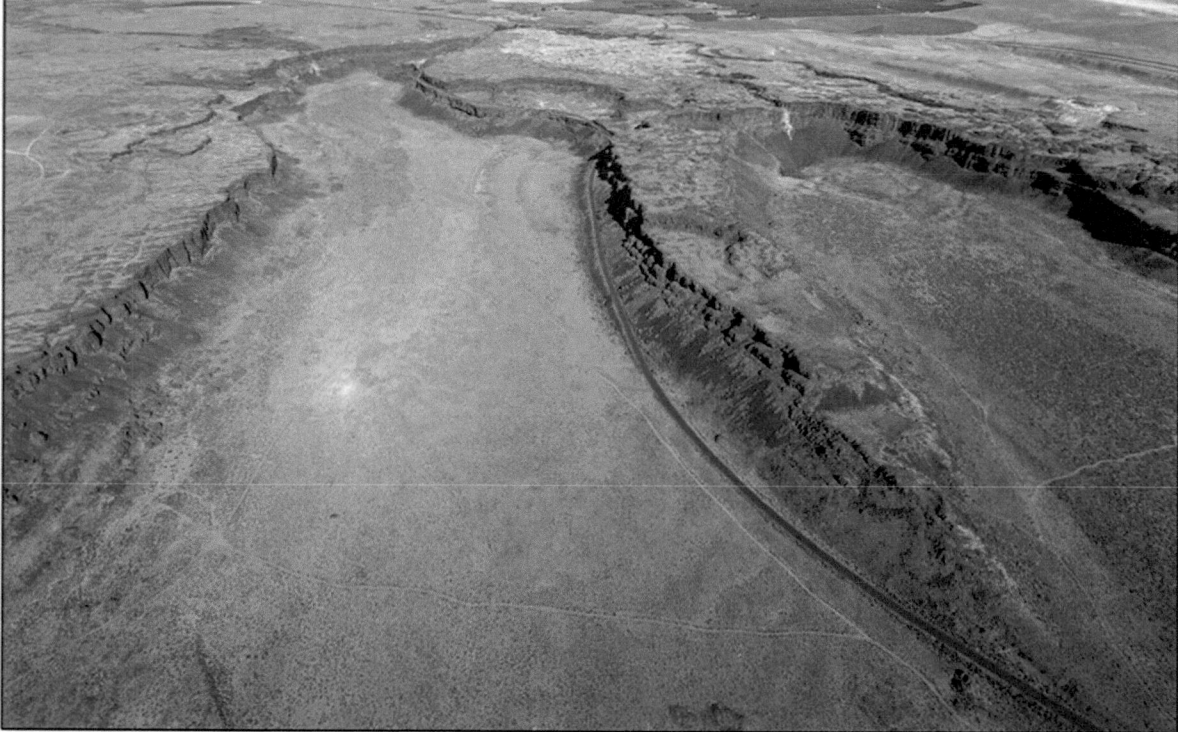

Frenchman Coulee Dual Recessional-Cataract Canyons Missoula floodwaters, many hundreds of feet deep, attacked Frenchman Coulee at the western margin of the Quincy Basin carving out the two canyons. A prominent rock blade, common of many recessional cataract canyons, separates the two amphitheater-shaped alcoves. The shorter canyon is known as Echo Basin.

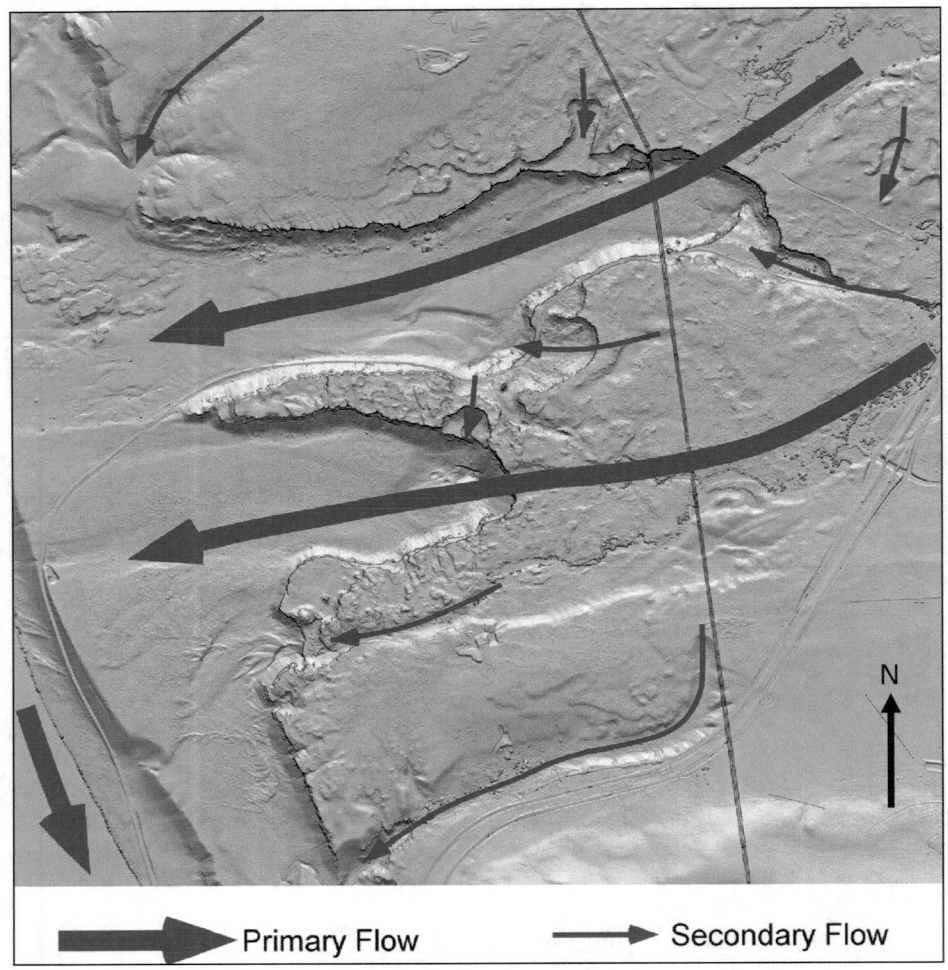

LiDAR Image of Frenchman Coulee The incredible detail of a LiDAR image shows the multiple different pathways for Ice Age floods into Frenchman Coulee. Basemap from the Washington Division of Natural Resources Lidar Portal (www.dnr.wa.gov/lidar).

Looking South Across Frenchman Coulee's Echo Basin

Potholes Coulee Another dual-cataract canyon, also with a distinctive rock blade down the middle, was carved out simultaneously with nearby Frenchman Coulee when voluminous floodwaters sought out and spilled over low points along the west side of the basin. Dusty Lake (in foreground) lies at base of one recessional cataract wall, while Ancient Lakes lie at head of the other canyon. Farmlands at upper right were generally above flood level and therefore retained their crop-producing soil. Looking northwest toward the Columbia River, hundreds of feet below, in the distance. Unusual landforms displayed on a topographic of Potholes Coulee caught Bretz's attention early on, leading him to begin a decades-long study of eastern Washington's scablands.

Dusty Lake Plungepool Basin A narrow blade of basalt divides the two parallel recessional cataracts. Dusty Lake and Quincy Lakes in the background. Looking east. Asterisk is location of the ice-rafted erratic shown on next page.

" 'The Potholes' are the best example mapped of a receding waterfall over lava flows known to the writer." Bretz (1923)

Potholes Coulee Erratic. A granitic ice-rafted erratic lies just upstream from the entrance into Potholes Coulee. Location is indicated by an asterisk on previous page.

Crater Coulee Crater Coulee was a third spillover out of the Quincy Basin into the deeper Columbia River valley. Here, a single straight and narrow flood channel suddenly widens and drops into a recessional cataract leading to a plunge pool. Looking southeast into direction of flood flow from the Quincy Basin. Preservation of soil cover on both sides of the coulee suggest floodwaters were restricted to the flood channel itself.

Babcock Bench Just west of the Quincy Basin lies an elevated surface (Babcock Bench) 600 feet above the Columbia River. Babcock Bench shows some of the most intensely eroded basalt scabland anywhere in the Channeled Scabland. Above, at left center, the scabland surface is covered with gravelly flood sediment, exposed within a borrow pit. Giant current ripples are preserved on both sides of the river, just beyond the borrow pit at left and atop West Bar at upper right. Crescent Bar, in the middle of the river, is a younger sediment bar lacking giant ripple marks. Looking south.

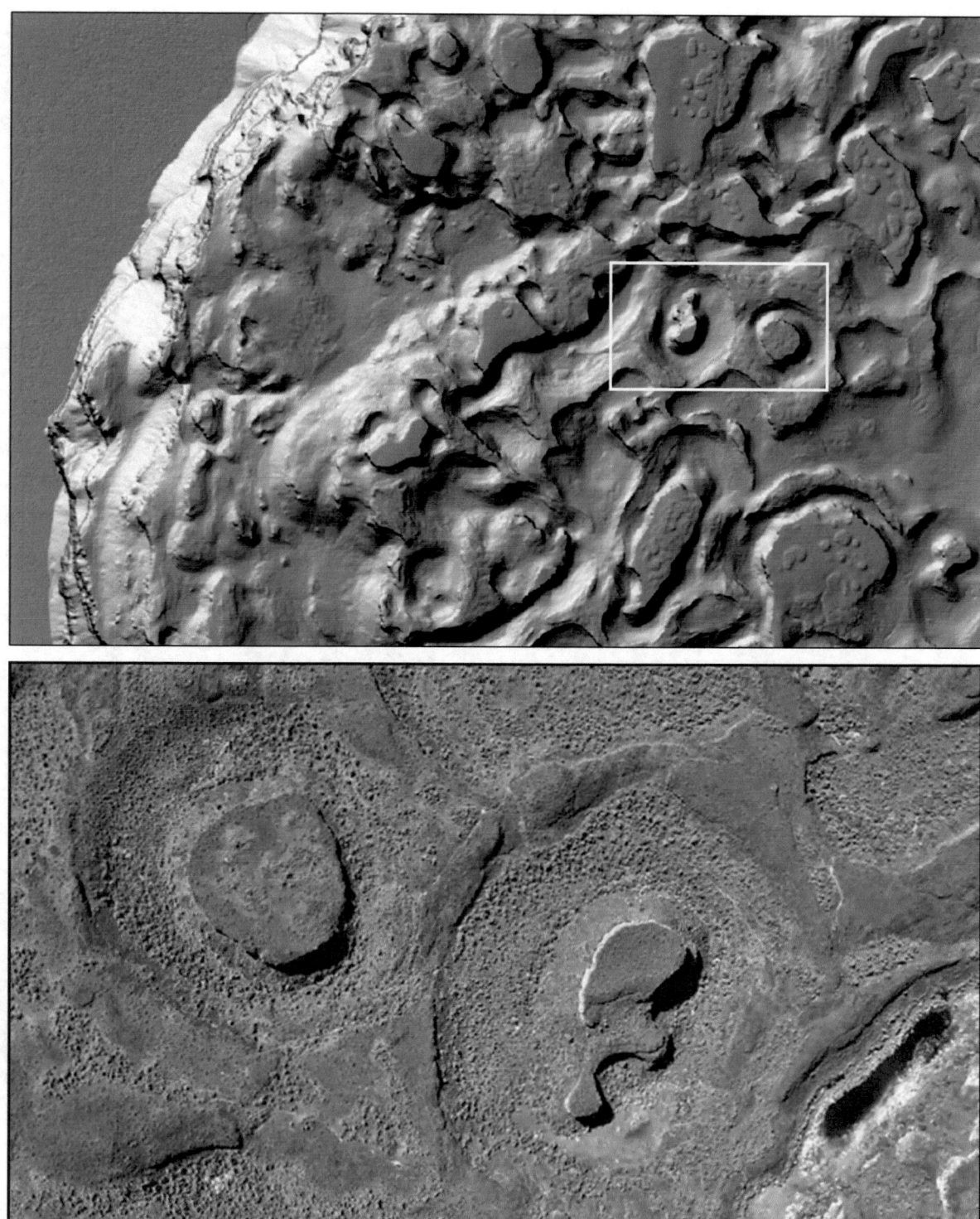

Erosional Chaos LiDAR image (above) enhances erosional patterns atop basalt bedrock on Babcock Bench. Irregularities in the hardness and structure of a basalt flow are reflected in the bizarre and unusual landforms atop the bench. From Washington Department of Natural Resources Lidar Portal (www.dnr.wa.gov/lidar). LiDAR, which stands for Light Detection and Ranging, is an exciting new remote-sensing technique that uses light in the form of pulsed lasers to measure variable distances to the Earth. An ordinary aerial photograph of the outlined area on the LiDAR image above is shown below.

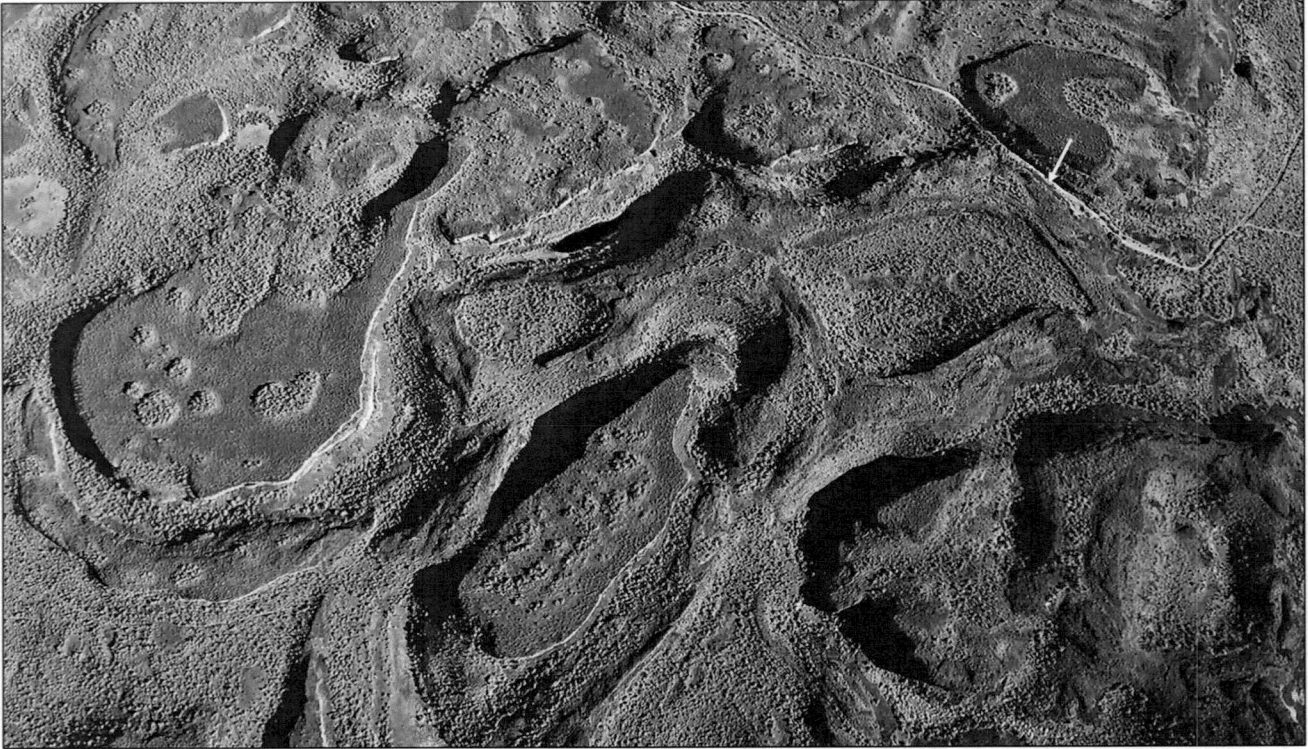

Bizarre Erosional Patterns atop Babcock Bench   Unimaginable megaflood erosion created this haphazard pattern of channels, buttes, mesas, moats and potholes. Abandoned jeep trail for scale (arrow in lower image).

Drumheller Channels A 12-mile wide swath of intensely eroded scabland developed here along a flow constriction at the southeast end of the Quincy Basin (see page 103), causing flood flow to temporarily increase due to the venturi effect. During the Ice Age the Columbia River (blue arrow) occupied the wide valley where the "underfit" lower Crab Creek flows today. It wasn't until the end of the Ice Age, with the breakup of the Okanogan Ice Lobe, that the Columbia River reestablished itself in its present location 30 miles to the west (see next page). Looking northeast.

"This spillway is a labyrinth of abandoned channels and rock basins among buttes, hills and higher scabland surfaces of basalt." Bretz (1928a)

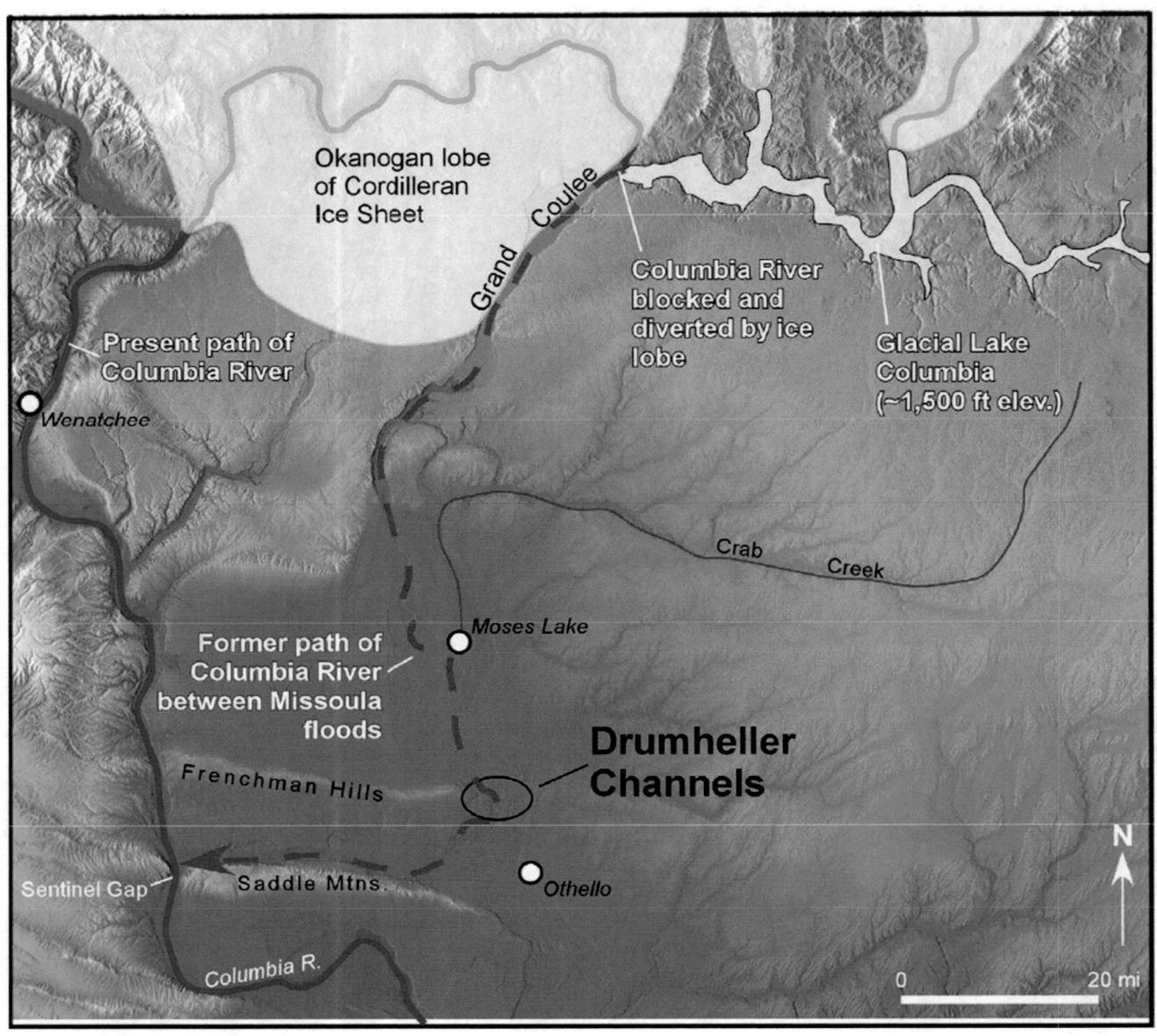

Shifting Columbia River During glacial times, not only did Okanogan Ice Lobe divert floodwaters out onto the Channeled Scabland, but also diverted the path of the Columbia River in between Missoula floods. As a result the Columbia River shifted to flow down Grand Coulee to present-day Moses Lake where it entered the Crab Creek drainage. From there it continued south through the Drumheller Channels before turning west toward Sentinel Gap. Here the river rejoined the present path of the Columbia River.

Streamlined Basalt Island Floodwaters flowed up to 300 ft deep over and around this tall eroded basalt upland within Drumheller Channels. Above, the island's prow points into the direction of flood flow. Some of the most classic flood-plucked columnar basalt is beautifully exposed around the perimeter of the island. The tall, polygonal columns formed along shrinkage cracks (below) as the lava flow slowly cooled and solidified after coming to rest here 10.5 million years ago.

Eagle Lakes Scabland Eagle Lakes occupy a portion of the flood-scoured Othello Channels. The Othello Channels developed, similarly to Drumheller Channels, as floodwaters funneled through an opening at the lower east end of the Saddle Mountains (see page 103). Notice giant grooves eroded into basalt bedrock that run parallel to the flood-flow direction similar to those in the Grand Coulee. Looking in the downflow direction.

Red Rock Coulee An especially long, cataract canyon near Royal City within the Othello Basin. The head of the recessional cataract (arrow) forms the Natural Corral—a horse-shoe shaped amphitheater. Lower Crab Creek Coulee (green area at upper right) and Red Rock Coulee were occupied simultaneously when megafloods drowned lower Crab Creek valley with hundreds of feet of floodwater.

Slackwater Basins

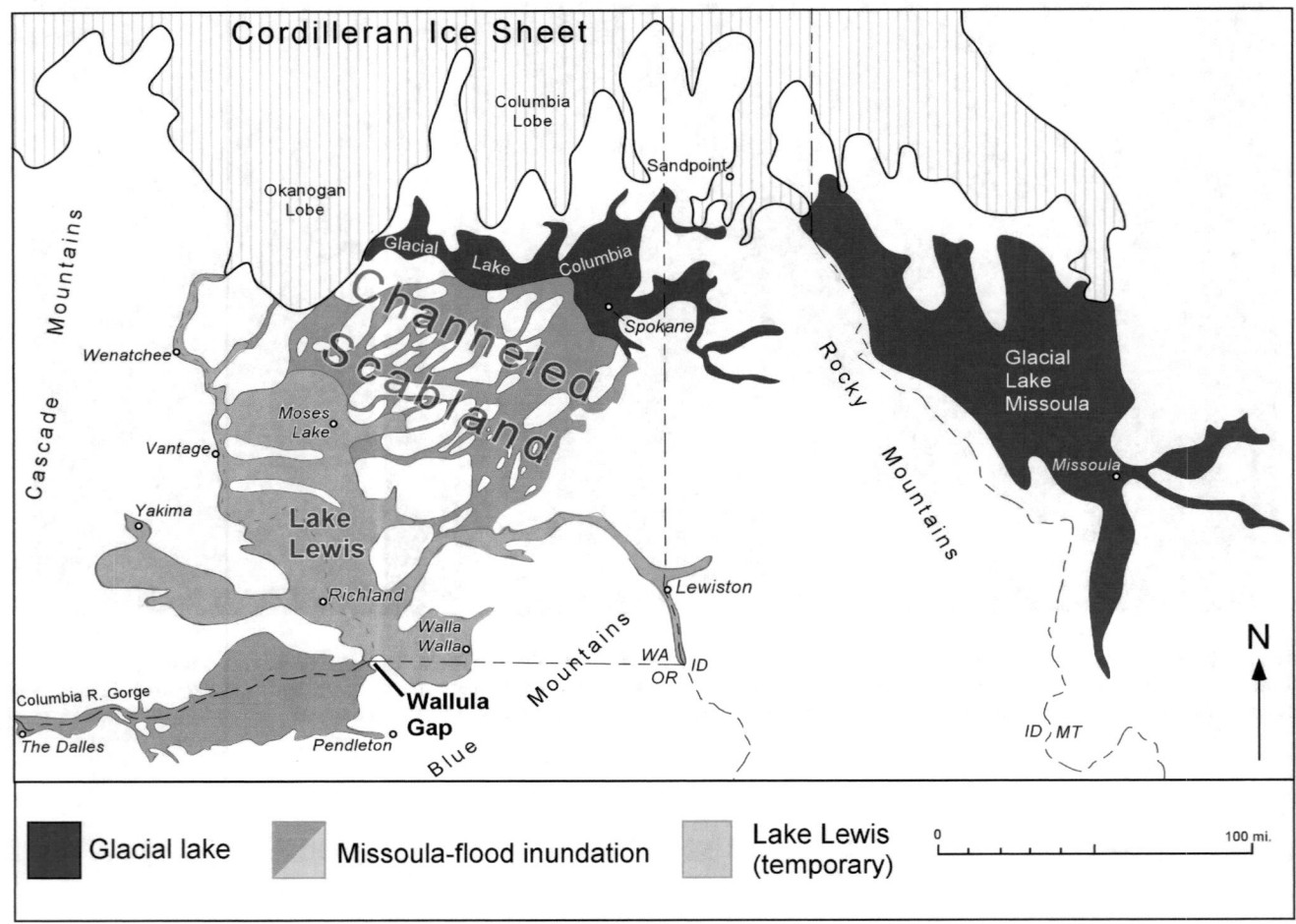

Backflooding Behind Wallula Gap After spreading out for up to 100 miles across the Channeled Scabland all outburst floodwaters from glacial Lake Missoula were forced through a single opening only a couple miles wide at Wallula Gap. Much more floodwater entered the gap than could pass through at once causing floodwaters to back up, creating temporary Lake Lewis. The backwater effect extended north to Moses Lake west to Wenatchee, and east to Lewiston, Idaho. Backflooding also reversed the flow of the Yakima and Walla Walla rivers.

Lake Lewis During the largest megafloods two to three times as much floodwater entered the Wallula Gap as could pass through at once, creating Lake Lewis (blue). The maximum height of Lake Lewis, ~1200–1250 ft elevation is based on maximum height of erosional scouring in Wallula Gap coincident with the maximum height of ice-rafted erratics observed within the Pasco Basin. Lake Lewis only lasted three weeks or less – the time it took for all the floodwaters to empty out to the Pacific. Flood currents also slowed way down and stagnated as floodwaters backed up into the Yakima and Walla Walla valleys, which were dead ends for the floodwater. At one time, during the largest Missoula floods, the backwaters of Lake Lewis contained about half the volume of Lake Missoula (~300 cubic miles). Downstream Lake Condon, another hydraulically dammed lake, backed up to ~1100 ft elevation behind another constriction (Rowena Gap) downstream in the Columbia River Gorge. Numbers refer to page numbers where features are described herein.

Lake Lewis Isles The largest Missoula flood that backed up behind a hydraulic constriction at Wallula Gap generated Lake Lewis – up to 900 ft deep. During this time only the very tops of nearby hills and ridges (e.g., Candy and Badger Mountains) rose above flood level creating the Lake Lewis Isles.

"A widespread submergence of the lower Columbia Valley is known to have occurred during the Wisconsin glaciation. It is recorded by berg-floated erratic boulders, some of great size, scattered widely in the Columbia Valley below the present altitude of about 1250 feet above tide." Bretz (1923)

"'Lake Lewis' is only a general term for possibly several Pleistocene pondings of the Columbia and its tributaries in the southwestern part of the plateau." Bretz, Smith and Neff (1956)

Lake Lewis Ice-Rafted Erratics   Thousands of exotic boulders came to rest in the backflooded basins associated with Lake Lewis when floodwaters temporarily ponded and slowed behind the Wallula-Gap constriction. The slowing currents allowed extra time for icebergs to drift and migrate to the margins of the basin before becoming grounded. Perhaps the highest concentration of erratics along the floods' path exists at Rattlesnake Mountain (above). Granitic boulders, like the one sitting here, make up about 75% of all the erratics mapped on Rattlesnake Mountain.

More Out-of-Place Erratic Boulders Above: angular boulder of banded argillite is perched atop the White Bluffs along the Columbia River. Below: A partially buried angular boulder of granodiorite rests peacefully in a vineyard, hundreds of feet above the Yakima Valley.

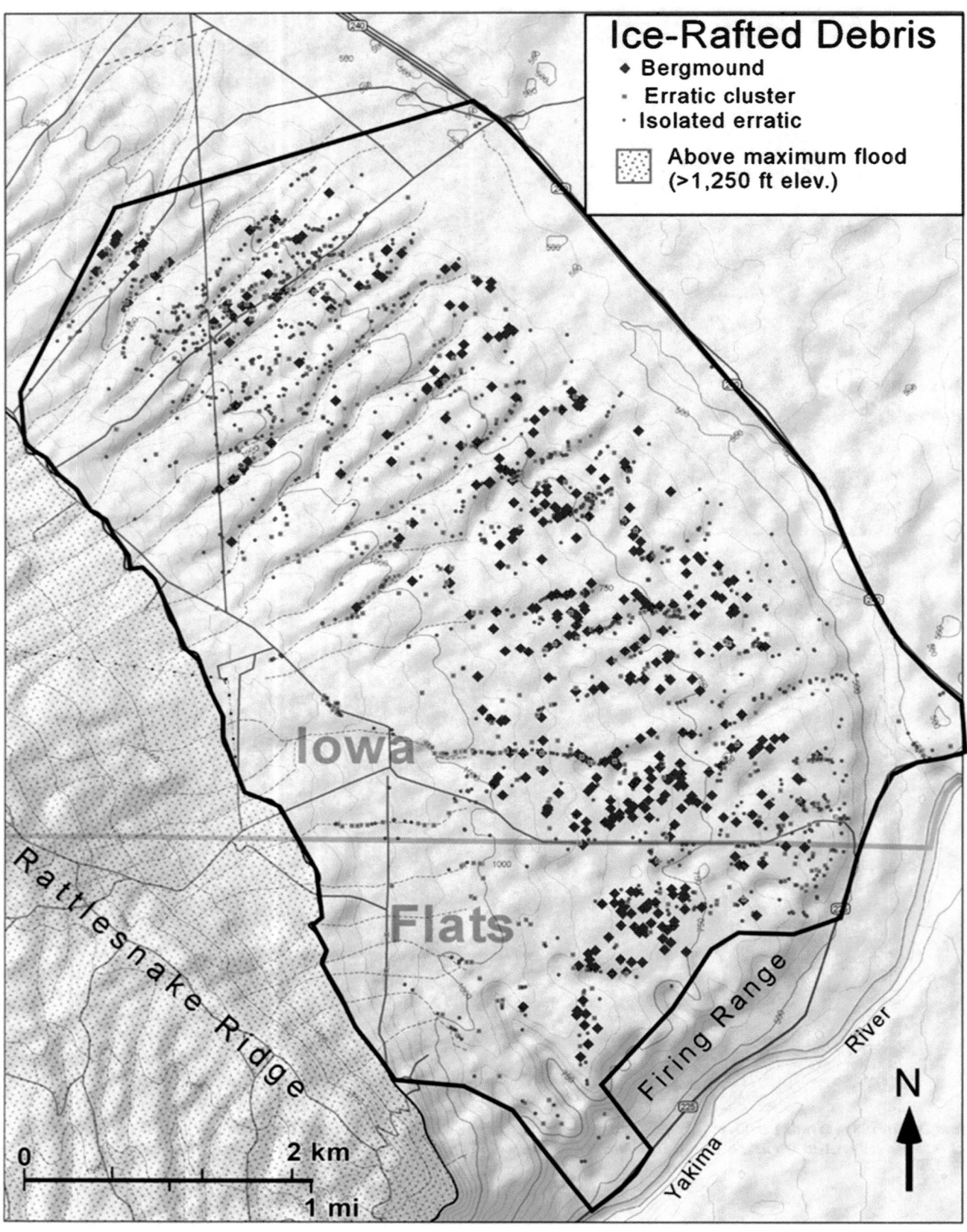

Erratic Behavior on Rattlesnake Mountain Represented on this map are over 2000 exotic erratic boulders and bergmounds that were mapped in a portion (23 sq mi) of Rattlesnake Mountain (Bjornstad 2014). Erratic clusters consist of two or more foreign boulders found in a tight grouping, probably rafted together in the same iceberg. The wide areal distribution is attributed to the large number of separate flood events, only a few of which were especially large.

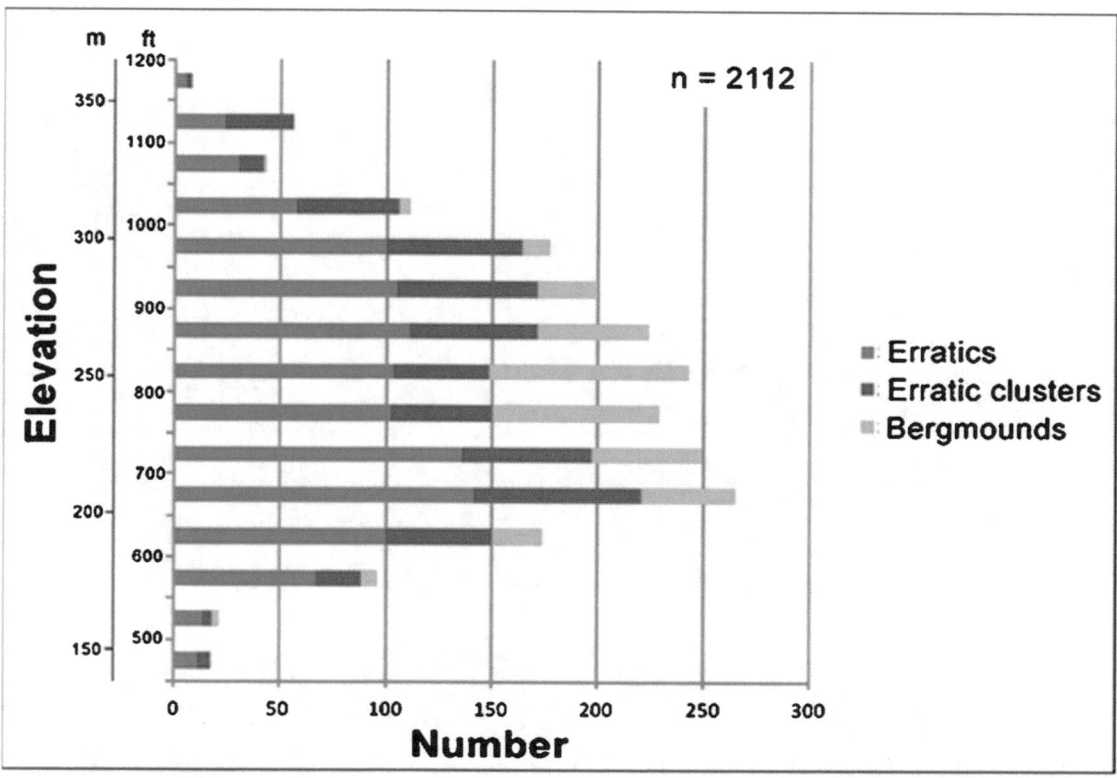

Vertical Distribution of Erratic Debris Note that the amount of erratic debris decreases with elevation to an upper limit of ~1200 ft elevation.

Boulder of Beautifully Banded Argillite on Rattlesnake Mountain This erratic likely came from the ice dam for glacial Lake Missoula more than 200 miles away. See also page 31.

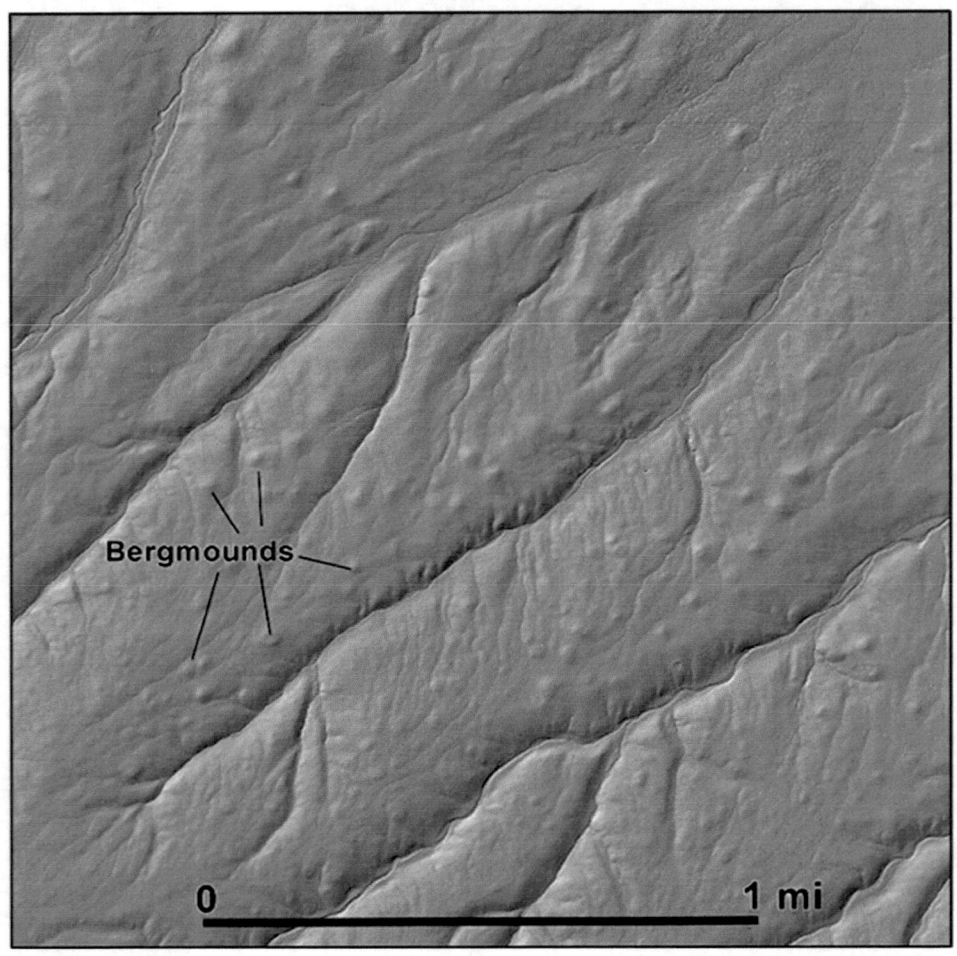

Bergmounds on Rattlesnake Mountain   Low-relief bergmounds were created via the melting of especially large and/or "dirty" icebergs containing lots of rafted sediment debris. Two bergmounds on the north flank of Rattlesnake Mountain (above). A single large ice-rafted granodiorite erratic (circled) rests atop left bergmound. Most bergmounds lie at an elevation below 1000 ft. Below, multiple bumpy bergmounds are clearly visible on LiDAR image from Rattlesnake Mountain.

"… isolated berg 'nests' and lone erratic boulders in so many places in the scabland, are irrefutable evidence for abundant floating glacial ice and, where closely spaced, are equally good evidence for the grounding of large numbers of these bergs. In Pasco basin, they obviously drifted ashore in a wide semiponded tract." Bretz, Smith and Neff (1956)

Sentinel Gap For millions of years, in the northwest corner of the Pasco Basin, the Columbia River has maintained a water gap as tectonic uplift slowly pushed up the basalt bedrock along the Saddle Mountains. The Columbia River here defines the western boundary for Ice Age floods that squeezed through Sentinel Gap. Looking downstream through the gap.

Mattawa Fan Immediately downstream of Sentinel Gap (arrow) is Mattawa Fan. Huge boulders of mostly basalt litter the surface for miles below the gap. Most of the debris was probably ripped from the sides of Sentinel Gap as megafloods reamed out and widened the gap. After racing through the gap flood currents slowed causing flood sediment to drop out of suspension. Occasional non-basalt boulders litter the fan as well – probably rafted here within melting icebergs. Looking north.

The Badlands (aka Chandler Narrows) Megafloods temporarily reversed the flow of the Yakima River (left) as they backflooded the valley. The narrow Badlands are located at the most-constricted portion of the valley. The initial pulse of floodwater moving through the constriction caused the water to speed up due to the venturi effect. Erosional scouring by the floods is indicated by scabland features like stripped-off mesas, rock benches and potholes within this constricted portion of the valley. Notice, in the distance, that productive farmlands reappear beyond the constriction where the flow of floodwaters slowed again – allowing for sediment deposition. Looking upvalley to the west.

" … this valley must have received a very marked back-rush into it from the main route of flood discharge… an advancing wall of water, like a gigantic bore, to sweep through the Narrows." Bretz (1930)

Slackwater Flood Rhythmites Graded, sedimentary layers of mostly sand and silt lie within backflooded basins and other areas where flood-waters slowed temporarily allowing for settling and deposition of suspended fine sediment. Here are two widely separated exposures of flood-rhythmites. The top image is from the northern White Bluffs—a slackwater area within the Pasco Basin along the Columbia River. Below is an exposure within the Walla Walla basin at Gardena Cliffs, 60 miles away. Multiple lines of evidence suggest individual rhythmites represent separate outburst floods from glacial Lake Missoula.

"The silt which characterizes the upper third of the Gardena cliffs is definitely arranged in strata of very fine gray sand grading up to into gray silt. The silt is commonly dense and hard, bending the edge of the tobacco tin into which a sample was scraped." Bretz (1929)

Burlingame Canyon Slackwater Rhythmites   Burlingame Canyon, in the backflooded Walla Walla Valley, reveals a record of at least 40 Ice Age floods during the last glacial cycle between 15 and 20 thousand years ago. A clean layer of volcanic ash (arrow), between two of the rhythmites, plus evidence for bioturbation between many of rhythmites suggest a period of time, several dozen years or more, separated individual flood events. See also page 20.

Origin of Burlingame Canyon   The 100-ft deep canyon was carved in only six days in 1926 at a breach in a nearby irrigation canal. In all, a total of 40 rhythmites are exposed in the steep canyon walls. This is the rosetta stone for the generally accepted multiple-megaflood hypothesis that originated in the 1980's. Public access to the canyon is closed and strictly enforced due to liability issues.

Rhythmite Detail Each rhythmite consists of a graded bed of mostly sand at the bottom transitioning to silt above. A sharp break exists between the tops and bottoms of individual beds. A thin layer of pure volcanic ash on the 12th rhythmite down (arrow) suggests there was a complete draining of Lake Lewis between the deposition of one bed and the next. The ash layer can be traced to an eruption of Mount St. Helens (set S) from ~16,000 years ago. Red bracket shows interval of detail for sedimentary peel at far right (Bjornstad 1980).

Megafloods and Fine Wine Some of the best agricultural lands in the Pacific Northwest are associated with slackwater flood deposits that accumulated within backflooded basins along megaflood paths. Shown here are expansive vineyards within the world-renowned Red Mountain Viticultural Area near the mouth of the backflooded Yakima Valley. Wine grapes are especially fruitful that grow in the fine topsoil derived from slackwater flood deposits within this and other backflooded valleys like the Walla Walla and Willamette Valleys, in addition to the Yakima Valley.

Wallula Gap

Hydraulic Constriction at Wallula Gap   Looking upstream where significantly more floodwater entered the gap than could pass through at once. The resulting constriction created temporary, hydraulically dammed Lake Lewis behind the gap. In this image floodwaters are ~1000 ft elev; the maximum flood level was another 200 ft deeper than illustrated here. Artwork by Stev Ominski.

Wallula Gap – East Side All outburst floodwaters from Glacial Lake Missoula were forced through this single opening before turning toward the Pacific via the Columbia River Gorge. Two Sisters, a local landmark, is shown silhouetted against the river.

"All Scabland drainage passed through this canyon and the flood reached 900 feet above present bottom..." Bretz, Smith and Neff (1956)

Two Sisters A pair of more flood–resistant pillars of 16-million-year-old basalt left standing since the last floods tore through the gap.

High Spillover Channels –West Side The largest megafloods overwhelmed the gap spilling over to carve scabland channels high above the gap along both sides.

Granitic Ice-Rafted Erratics along West Side of Wallula Gap Boulder in upper photo is 700 ft, and lower photo 800 ft, above the Columbia River.

E. Columbia River Gorge

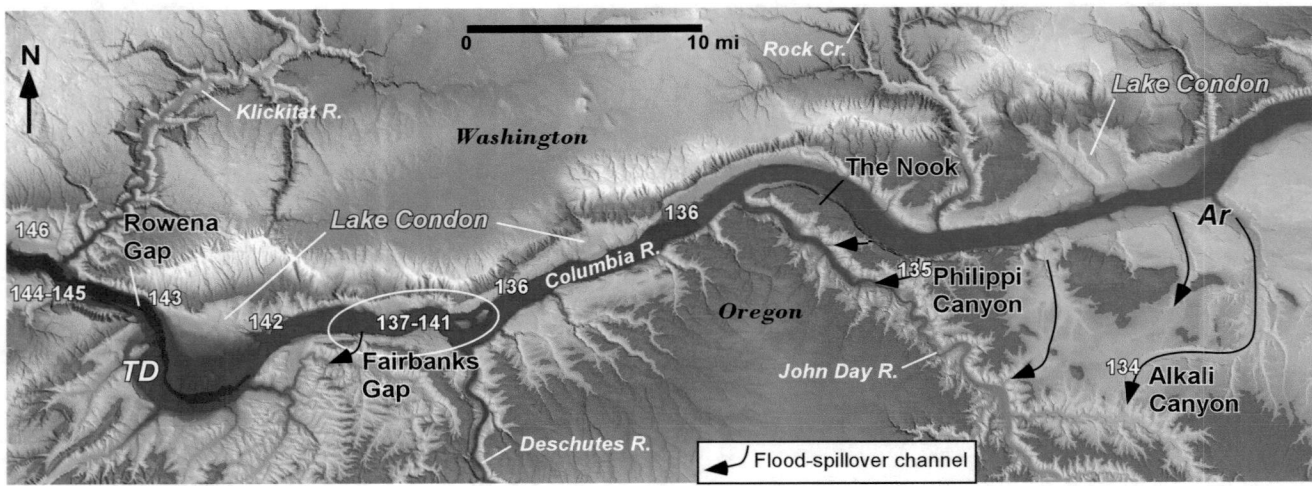

Eastern Columbia River Gorge Index Map Maximum floodwater level (~1100 ft elev.) for Lake Condon in blue. Lake Condon formed when megafloods backed up behind a hydraulic constriction within the Gorge at Rowena Gap. Numbers correspond to page numbers herein. Ar = Arlington, TD = The Dalles. Numbers refer to pages of flood features described herein.

Alkali Canyon Coulee Detour After exiting Wallula Gap megafloods headed west for the Columbia River Gorge. The broad, curving megaflood channel marks a spillover route for megafloods that temporarily separated from the Columbia Valley at Arlington, Oregon in the eastern Gorge. Floodwaters that took this circuitous route emptied into the John Day River Canyon several miles downstream, before eventually rejoining the Columbia River 20 miles further downstream.

High Spillover at Philippi Canyon (aka "The Narrows") As the largest Ice Age megafloods, up to 800 feet deep, raced down the eastern Columbia River Gorge some of the floodwater spilled over, via Philippi Canyon, into the adjacent John Day River canyon (visible in background). Other floodwaters that flowed down John Day River via Alkali Canyon joined those of Philippi Canyon ~10 miles upstream along the John Day River. I-84 at bottom follows the entire length of the Oregon side of the Columbia River Gorge. Looking south.

Hidden Philippi Scabland As floodwaters from the Columbia valley spilled over from the lower right they swung around through this elevated, incised scabland channel before diving into the John Day River valley far below at upper left. Simultaneously, floodwaters from Alkali Canyon Coulee raced down the John Day valley from the left— starting 10 miles upstream.

"Only a brief life for The Narrows spillover could have been possible before backflooding up the John Day brought it to an end." Bretz (1969)

Flood-Swept Rock Bench at Maryhill   The Maryhill Museum of Art is perched high above the Gorge upon a basalt bench 600 feet above the river. Another 200 ft of water would have submerged the museum if it existed during the largest megaflood. Looking upriver.

Chaotic Spindly Basalt Erosion   Some unusual flood erosion created hoodoos here upon a brecciated flowtop of flat-lying Grande Ronde Basalt in the eastern Columbia River Gorge.

Miller Island Trenched Spur The primary channel for the Columbia River is on the left side of Miller Island above. During megafloods the extremely fast and powerful currents forced floodwaters to flow in a straighter line up and over the top of the basalt spur (arrow). The end result was a more direct route that eroded a new channel across the spur. Looking downriver towards Oregon's volcanic Mount Hood (west).

Flood-Swept Celilo Bench Here lies a broad denuded rock bench above the drowned Celilo Falls. Looking up the Columbia River into the direction of the megafloods. Arrow points to the Miller Island trenched spur featured above. The Deschutes River joins the Columbia River at upper right.

Celilo Scabland It's amazing that most drivers travelling through the Columbia River Gorge via I-84 are totally oblivious to the wild erosional scablands behind cliffs immediately above them near Celilo. Three distinct rock benches were created as megafloods successively eroded and peeled back several basalt flows like layers of an onion. Linear grooves on the basalt surface developed where megafloods eroded away weaker rock along pre-existing tectonic fractures. Maximum flood level is indicated by the uppermost planar erosional escarpment (arrow above) – similar to faceted escarpments within the Channeled Scabland. Deschutes River enters the Gorge at upper left. Below: several circular potholes are eroded onto the upper basalt bench. Looking southeast.

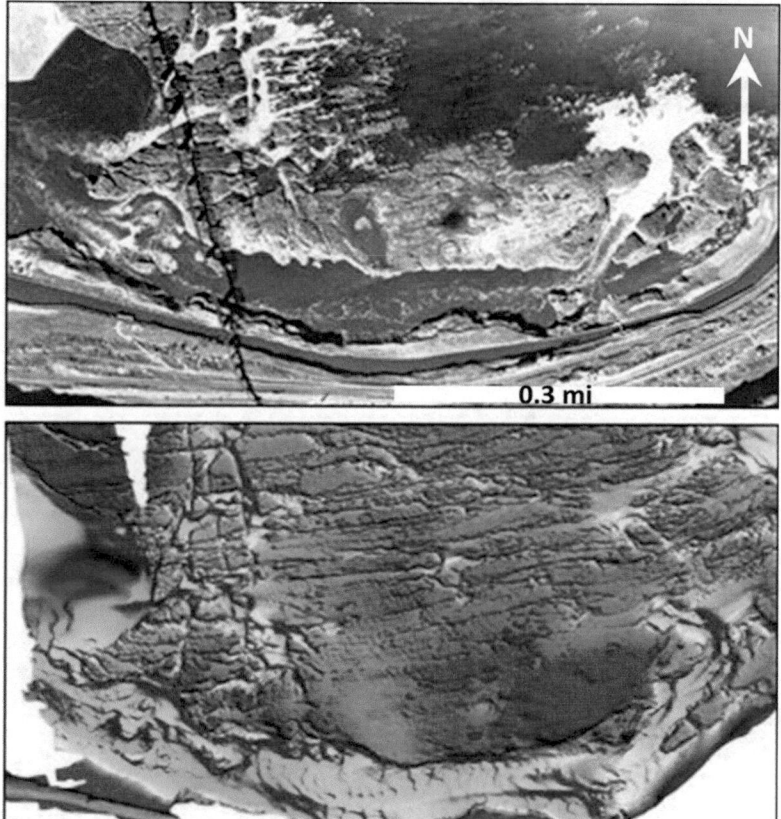

Celilo Falls   Top: Fishing grounds prior to the damming of the Columbia River behind The Dalles Dam in 1957. Below: two aerial images of the Celilo Falls area; the upper black and white image was photographed before the man-made inundation of the river while the color image is from a 2007 sonar-reflection survey obtained, remotely, from the lake surface created by the dam. Both images show the deeply scoured and eroded basalt surface created by Ice Age floods. The color image, in particular, clearly reveals a cataract canyon that migrated eastward during recession (lower-elevation green area along bottom of lower image). Heavily flood-grooved basalt surface (in red) is also striking and very similar to grooved surfaces described previously near Dry Falls (see page 14). For reference, the railroad bridge in the upper photo is the same bridge in photo on next page. Above images courtesy of the U.S. Corps of Engineers.

Celilo Falls

Lake Celilo and Fairbanks Gap   The Columbia River flows beneath a railroad bridge just downstream from where Celilo Falls once cascaded — now part of Lake Celilo created by the man-made Dalles Dam. During megafloods, exceedingly deep and fast-moving floodwaters escaped out of the Columbia valley carving a prominent spillover channel through Fairbanks Gap (arrow). Looking west toward Oregon's majestic Mount Hood.

Fairbanks Gap Spillover Megafloods >800 ft deep spilled over from the Columbia Gorge through Fairbanks Gap into an adjacent drainage. Top: Triangular escarpments (arrow) on both sides of the channel were eroded and planed off in post-basalt sediments by the escaping floodwaters. The height at the base of the spillover channel lies 550' above the Columbia River far below. Below: looking into the gap from the opposite side toward the Columbia Gorge. Arrow indicates flood-flow direction.

Line of Flood-Faceted Escarpments Along the Washington side of the Columbia Gorge, near Wishram, extremely deep, high-velocity flood-waters trimmed off the valley wall, leaving behind a line of faceted rock spurs (arrows) in their wake. Note the flood-swept, scabland-like bench below the escarpment. Tip of Mount Adams volcano at top center.

Horsethief Butte Scabland At Horsethief Butte State Park Ice Age megafloods stripped away all loose sediment below 1000 ft elevation in the eastern Columbia River Gorge. Horsethief Butte (center left) is an erosional remnant of a once continuous basalt flow. WA State Route 14 at right follows one of many megaflood channels. Looking southwest toward The Dalles. After spreading out into the Dalles Basin outburst floodwaters converged onto another major hydraulic constriction at Rowena Gap (red arrow).

W. Columbia River Gorge

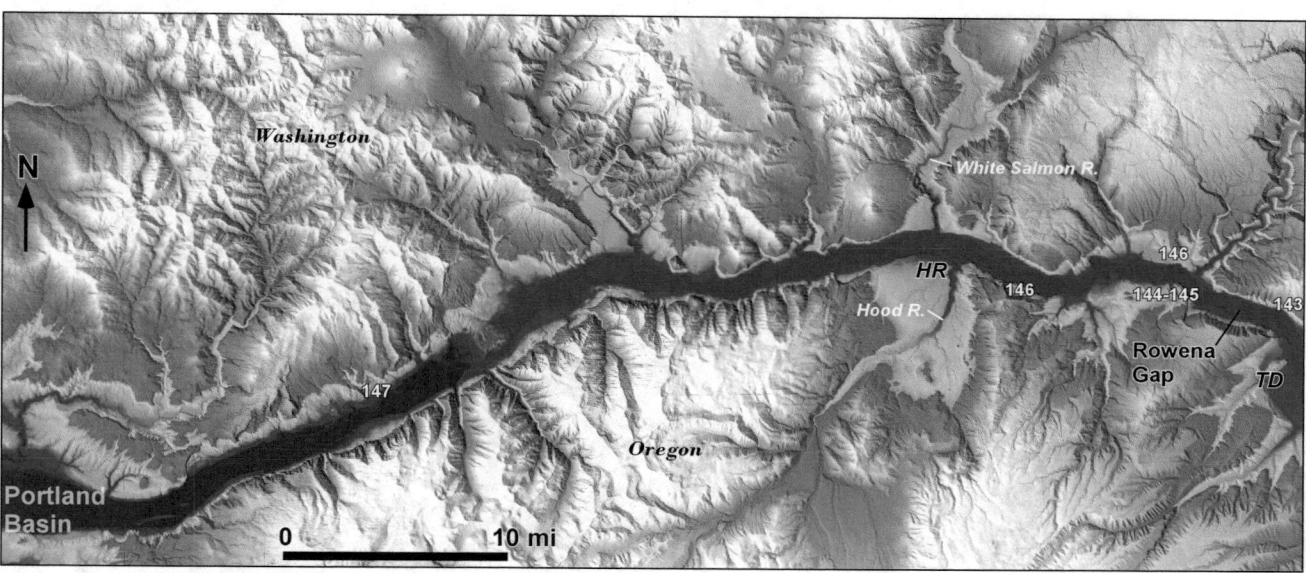

Western Columbia River Gorge Index Map While there was a steady downstream gradient for the maximum megaflood through the western Columbia Gorge, represented here is an average value (~720 ft elev.) shown in blue. Numbers correspond to page numbers for flood features described herein HR = Hood River.

Megaflood Strandlines at Rowena Gap High-water marks (arrow) for Ice Age megafloods are visible at Rowena Gap where flood-eroded sediment covering the basalt bedrock allowed for preservation of ancient flood levels within the Gorge.

Rowena Gap Constriction   Looking east (upriver) the Columbia River cuts through a high ridge of upfolded basalt (Ortley Anticline) at Rowena Gap. This narrow opening (red arrow) along the Columbia Gorge acted as another hydraulic constriction when the Gorge was overwhelmed by megafloods causing floodwaters to temporarily pond up to 1100 ft elevation (1000 ft deep!) behind the gap. Speeds of megafloods at constrictions like Rowena Gap may have approached 80 mph through the Columbia River Gorge here. Downstream of the gap, the still-powerful floodwaters gouged out deep canyons and mesas at the Rowena Dells in the foreground.

Rowena Dells   Ice Age megafloods (arrow), flowed up to 500 ft deep above the elevated, flood-swept plateau at Rowena Dells. Here is a plethora of precipitous canyons, rock benches, and potholes eroded along the Oregon side of the Columbia Gorge. Rowena Crest Viewpoint is at lower right and Washington's Mount Adams volcano in the distance.

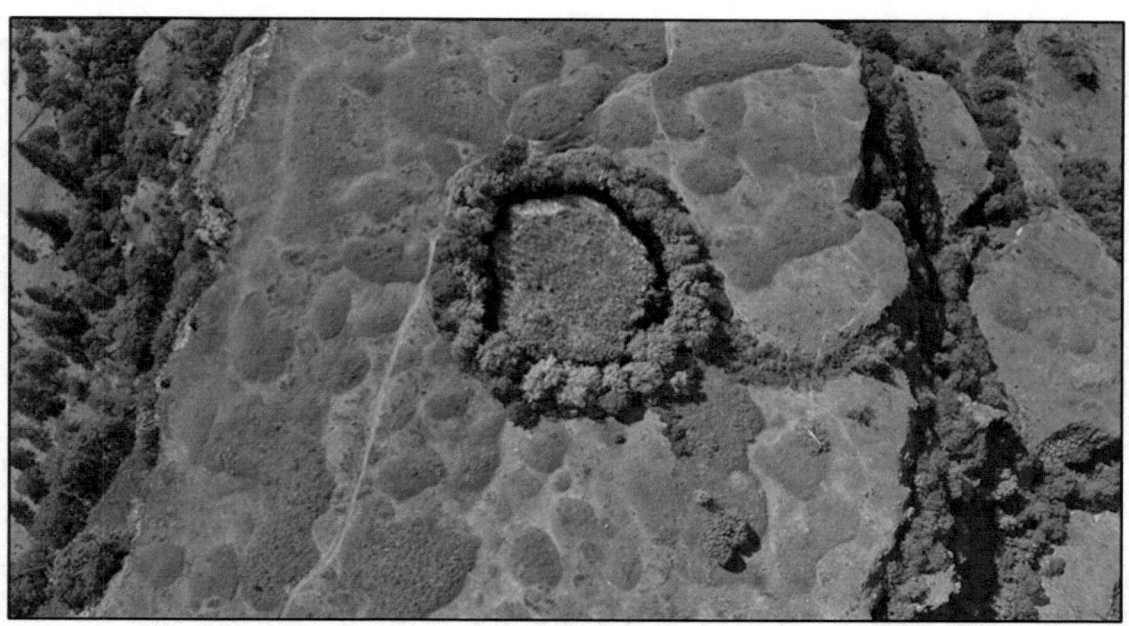

Elevated Flood-Swept Plateaus, Canyons and Potholes at Rowena Dells Several giant potholes excavated by deep swirling floodwaters (kolks), stretch out along a flood-swept rock bench at the Rowena Dells. Above is one of the potholes – today a tree-lined boggy swamp. Notice foot path for scale. Bumpy surface is patterned ground that has developed in surficial, fine-textured sediment since the Ice Age.

Flood-Scrubbed Bedrock Megafloods raced through the Columbia Gorge from right to left across this sloping surface (i.e., tectonically folded basalt of the Mosier Syncline) a few miles west of Lyle along the Washington side of the Gorge. Dashed lines mark the trimline for the floods based on the upper limit of soil erosion by megafloods. This upper limit of scrubbing was about 700 ft above Columbia River (lower left corner). Washington State Route 14 at lower left.

"Scrubbed basalt slopes and cliffs are prominent features on both sides of the river above Lyle." Bretz (1925)

Short Recessional-Cataract Canyon along Hatfield Trail An unusually short (0.5 mi) recessional cataract canyon eroded in basalt along the Oregon side of the Columbia Gorge across the river from Bingen, WA. The floor of the coulee is relatively flat except at the headwall and the mouth, which hangs 325 feet above the Columbia River. Visible along the shadowed floor of the coulee are a couple of elongated, flood-scoured ponds (arrows). Today, Oregon's Mark O. Hatfield Memorial Trail passes by the head of this rather unique coulee.

Beacon Rock   Beacon Rock is the erosional remnant of a Cascade volcano that erupted ~57k years ago. During megafloods, which approached the summit, most of the volcano was eroded away except for the more-resistant interior neck of the volcano. Today, one can access the pinnacle's summit via a thrilling zig-zag hike that climbs 850 ft via 53 switchbacks clinging to it's near-vertical sides (arrow).

Megaflood Rips Through the Lower Gorge   Crown Point, a local landmark, today rests on bench in right foreground. Top of Beacon Rock (arrow) pokes out above floodwaters. Painting "Ages End" by artist Stev Ominski. Looking east

Lake Allison

Temporary Lake Allison After exiting the Columbia River Gorge megafloods immediately spread out into the Portland Basin. Further down-stream outburst floodwaters backed up behind another hydraulic constriction at Kalama Gap, forcing floodwaters south into the Willamette Valley via two flood channels at Lake Oswego and Oregon City (arrows). Soon after the flow of water drained out through these same narrow conduits. The backwater rose to a maximum of 400 ft elevation creating Lake Allison, which likely only lasted a couple of weeks or less before all floodwaters had drained out to the Pacific. Po = Portland, Sa = Salem, Eu = Eugene

Willamette Valley's Bellevue Erratic   Today, a short trail leads to the 90-ton argillite boulder within the Erratic Rock State Natural Area near Bellevue, Oregon. Below: how the ice-rafted Bellevue Erratic might have appeared >15,000 years ago, as the cluster of exotic boulders melted out of a grounded iceberg soon after Lake Allison drained from the Willamette Valley.

Tip of the Iceberg   A re-creation of what Bellevue Erratic might have looked like soon after grounding along the margins of Lake Allison. Painting by Stev Ominski.

Aerial of Bellevue Erratic   The Bellevue Erratic (circled) rests upon a gentle hill 150 ft above the floor of Oregon's bucolic Willamette Valley. The erratic floated in on an iceberg during a megaflood that backfilled the valley, creating temporary Lake Allison (backwater behind Kalama Gap constriction). This erratic lies at about 300 ft elevation or about 100 ft below the maximum flood level for the valley. The Bellevue erratic is composed of an angular and tabular block of banded argillite – identical to other erratics from the Belt Supergroup – traceable to the ice dam for glacial Lake Missoula hundreds of miles away in northern Idaho. Along with the large argillite erratic are a few smaller pieces of similar rock nearby – probably all transported together within the same iceberg.

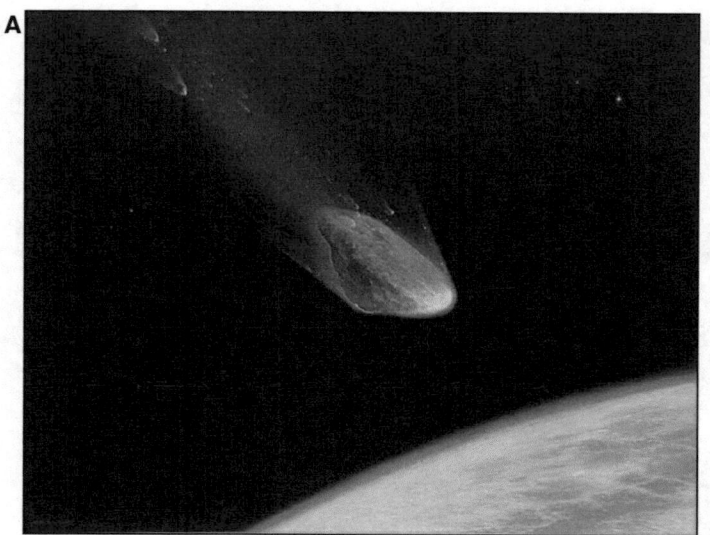

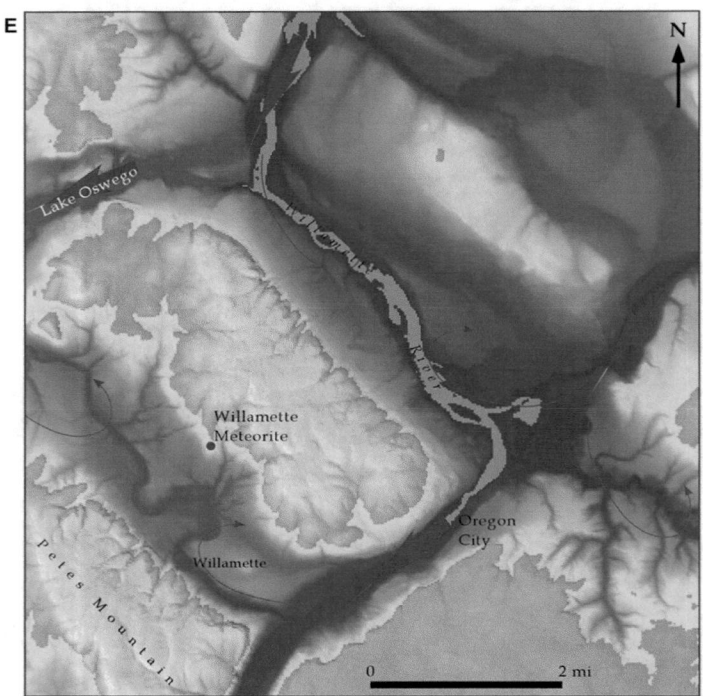

The Willamette Meteorite—A Most Unusual Ice-Rafted Erratic   The largest meteorite yet discovered in U.S. was discovered upon the surface of a steep hill just south of Portland, Oregon near the local high-water mark (400 ft elev.) for Ice Age megafloods shown in the map (**E**) above. From this map it's easy to visualize how circulating back eddies during megafloods would have naturally pushed floating icebergs and their estranged payloads toward the location of the discovered meteorite, as floodwaters rushed southward to fill the Willamette Valley. First discovered by a white settler in 1902, the meteorite had long before been revered by local Native Americans. The lack of any evidence of an impact crater, but being in proximity to other foreign erratic boulders, strongly suggests this iron-nickel meteorite rafted into place, along with other erratics, on an iceberg during one of the largest Missoula floods ~17 to 18k years ago. The streamlined shape of the meteorite (**C** and **D**) appears to be inherited from its descent through the earth's atmosphere (**A**) before slamming into the Cordilleran Ice Sheet—probably somewhere in today's British Columbia. As the ice sheet crept southward, eventually the ice-entombed meteorite became part of the ice dam for glacial Lake Missoula. Upon breakup of the ice dam it was rafted within a floating iceberg (**B**) hundreds of miles away to the Portland area. The deeply pitted surface (**C** and **D**) is the result of corrosion by western Oregon's highly acidic soils that have since ate away at the metallic meteorite. Since 1906 the 15.5 ton Willamette Meteorite has been on display at New York's American Museum of Natural History (**D**). Re-creations of the meteorite in A and B above are by artist Stev Ominski.

Glacial Lake Columbia Flood

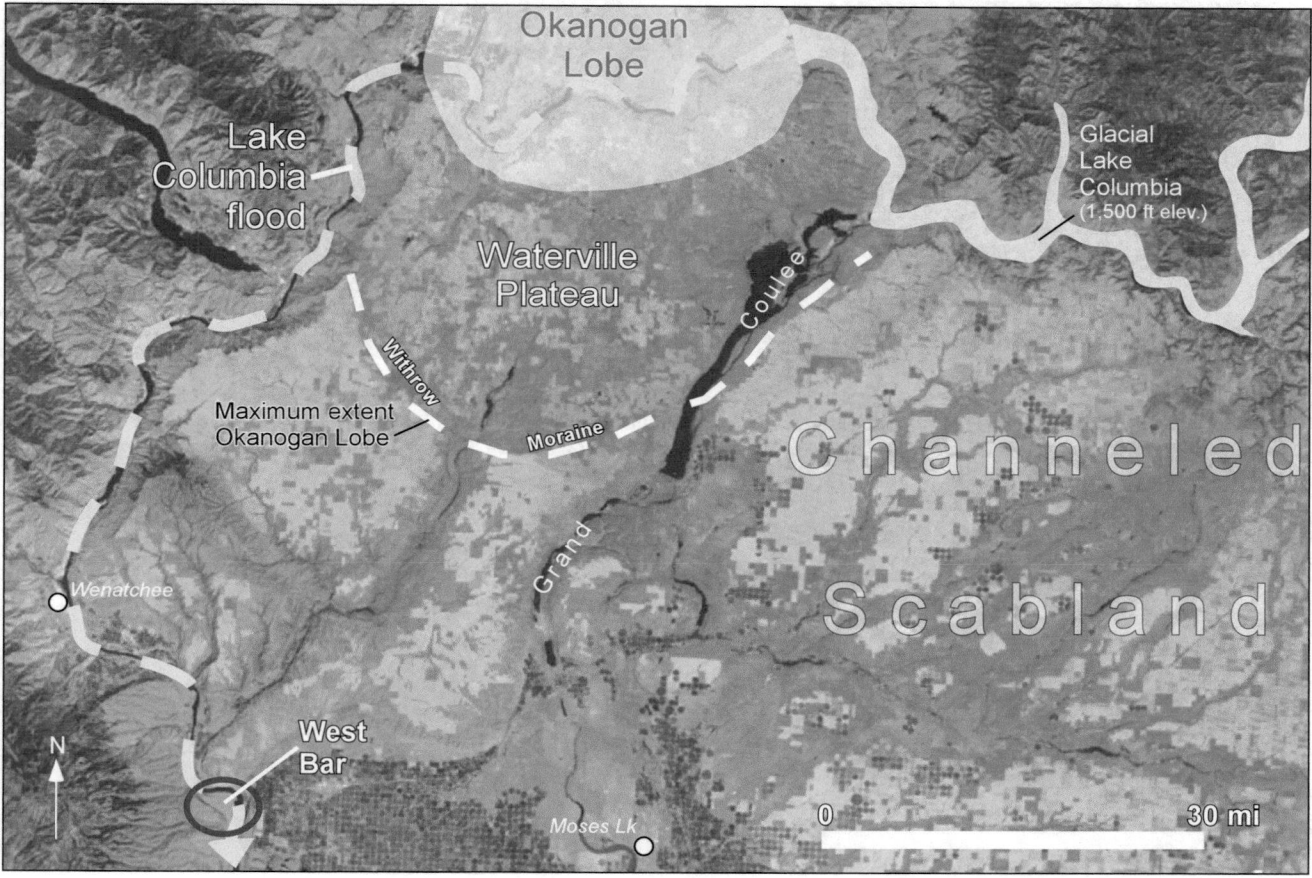

Pathway for the Lake Columbia Outburst Flood Several hundred years after the last flood from glacial lake Missoula, there was at least one last flood that occurred during the breakup of the Okanogan Ice Lobe. The Okanogan Lobe retreated northward from the Waterville Plateau at the end of the Ice Age about 14,000 years ago. That's when an outburst flood—probably from the sudden breakup of the glacial Lake Columbia and restricted to the Columbia River valley, flowed west through the breached ice dam before turning south toward West Bar (dashed blue line).

© Springer Nature Switzerland AG 2021

B. N. Bjornstad, *Ice Age Floodscapes of the Pacific Northwest*, https://doi.org/10.1007/978-3-030-53043-3_4

West Bar Giant Current Ripples It is clear, based on the orientation of giant ripples on West Bar, that the flood current coming down the Columbia Valley from the north—and not from one of the last Missoula floods, was responsible for the giant ripples preserved at West Bar. The orientation of the ripples would be very much different if coming from a Missoula flood. Looking downstream (south).

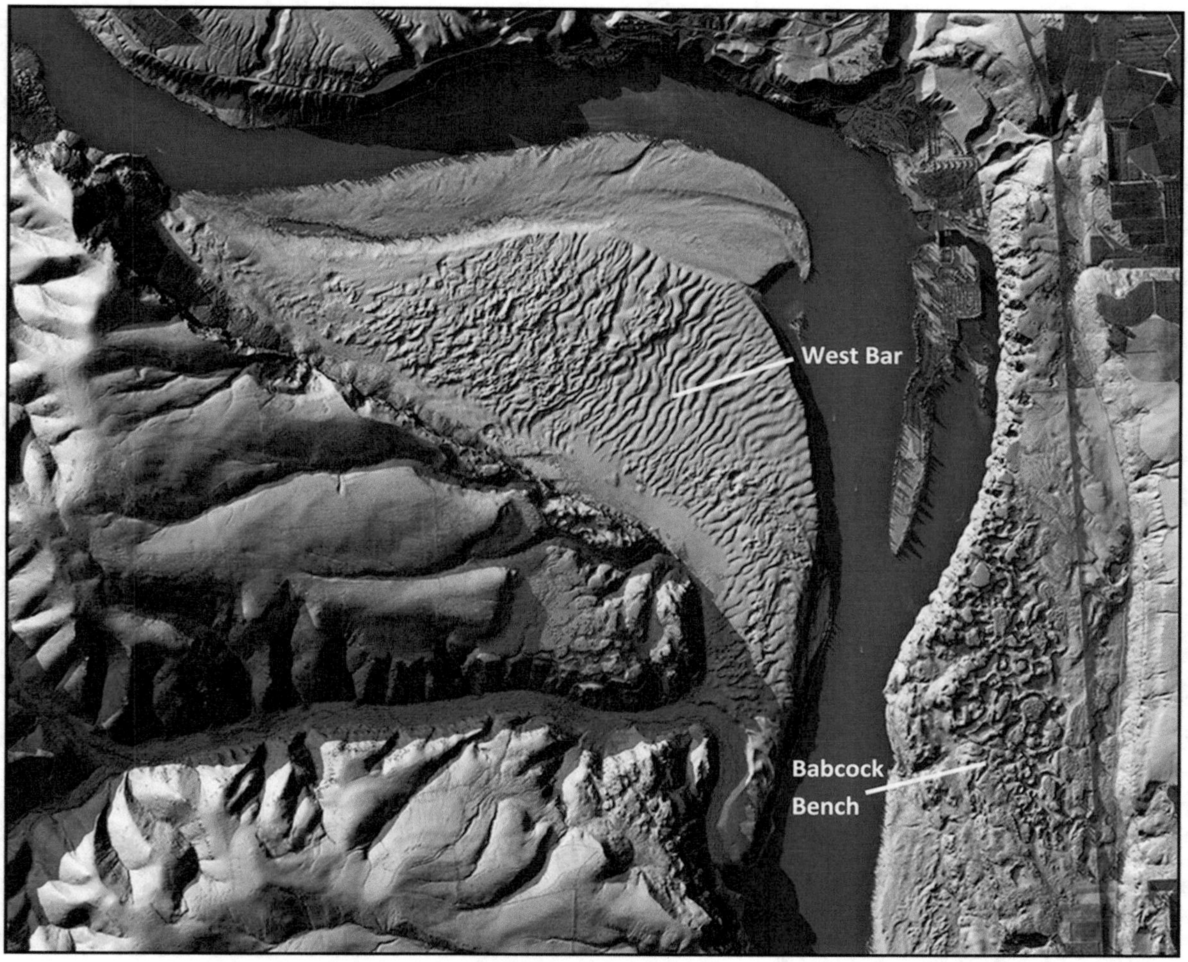

West Bar Giant Current Ripples Via LiDAR Imaging LiDAR imaging (above) from the Washington Department of Natural Resources brings out amazing topographic details on the surface of West Bar. Also displayed is incredibly chaotic megaflood erosion of basalt surface atop Babcock Bench (see pages 110–112). While the West Bar ripples developed during deposition from the last Pleistocene flood the erosion on Babcock Bench occurred much earlier during one or more of the largest Pleistocene floods, presumably from glacial Lake Missoula - before the Okanogan Lobe blocked the Columbia River.

"Seen from viewpoints along the highway east of the river, the surface of West Bar seems to be marked by great current ripples." Bretz (1930)

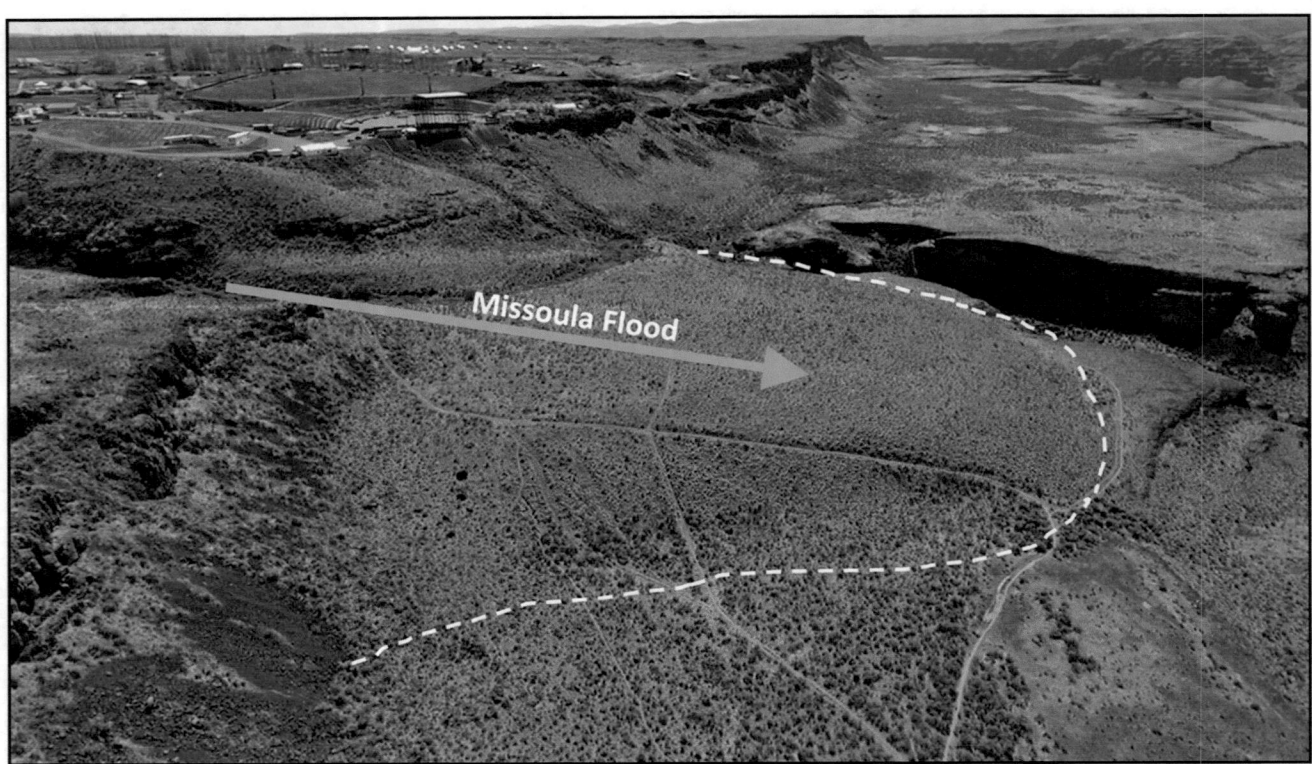

Upper Limit for Lake Columbia Flood A well-preserved delta-like flood bar (dashed) protrudes out onto Babcock Bench 400 ft above the Columbia River (upper right). The bar originated during one of the larger, early Missoula floods as floodwater spilled over (arrow) a divide along Evergreen Ridge out of the Quincy Basin and onto Babcock Bench. Much later, around 14,000 years ago, the glacial Lake Columbia flood drained through the valley below. Apparently, the volume of the Lake Columbia flood was insufficient to overtop Babcock Bench, since there is no evidence of flood erosion to the delta bar. For example, the nose of the bar would have been easily truncated or trimmed off if a younger flood came in contact. Therefore, the maximum depth for the Lake Columbia flood here is limited to <400 ft or <980 ft elevation. A popular music venue, the Gorge Amphitheater, is at upper left. Looking south.

East Wenatchee Boulder Bar The surface of a low-elevation bar along the Columbia River is covered with huge rounded boulders, mostly of granitic composition, like this one. While the last Missoula floods likely backflooded UP the Columbia River to here, there is no logical way boulders this size could have been transported via backflooding or from normal flows of the Columbia River. The most reasonable explanation is they were carried downstream during the Lake Columbia flood ~14,000 years ago.

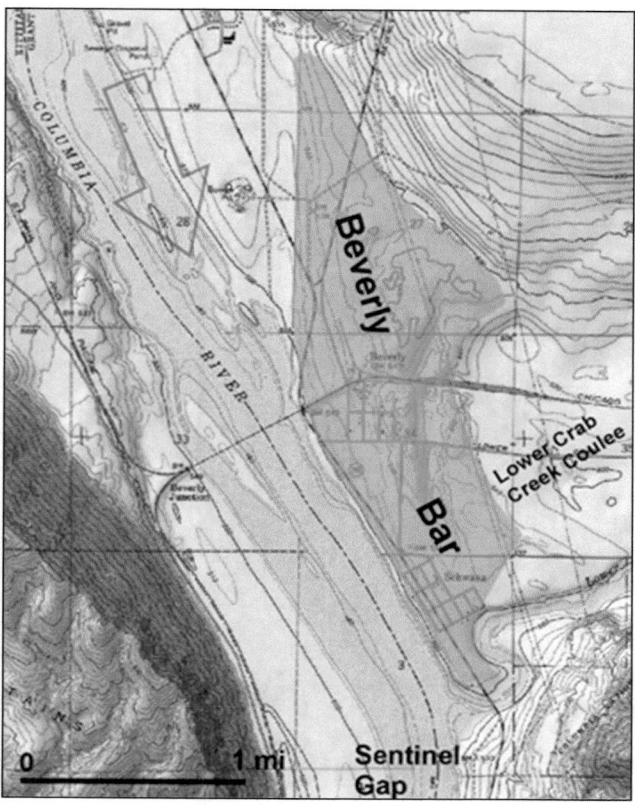

Beverly Bar Far downstream from glacial Lake Columbia, just north of Sentinel Gap at Beverly, lies a telling flood bar. This bar lies across the mouth of lower Crab Creek Coulee, which had previously conveyed Lake Missoula outbursts. The orientation and elongation of Beverly Bar clearly indicates it is from a last flood (arrow) coming down the Columbia River valley from the north (i.e., Lake Columbia). The bar could not have formed during a previous Missoula flood since floodwater flowing down lower Crab Creek from the east would have easily eroded through or destroyed this bar.

Lake Bonneville Flood

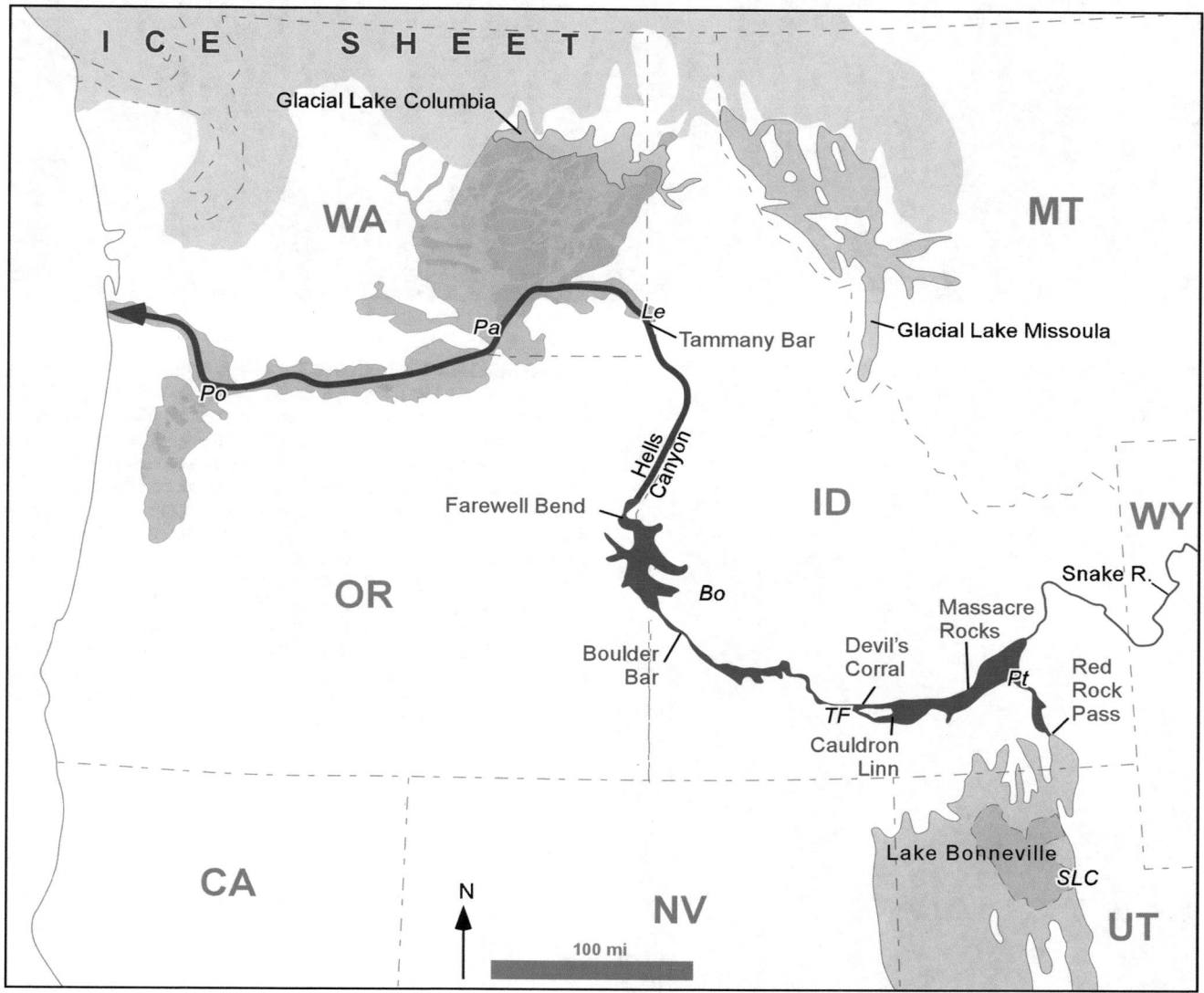

Pathway for the Lake Bonneville Flood The Lake Bonneville flood started in the extreme southeast corner of Idaho at Red Rock Pass and followed a relatively narrow path most of the way to the Pacific Ocean (O'Connor 1993). After spilling through Red Rock Pass the flood joined the Snake River canyon near Pocatello. From there, the flood generally followed the Snake River canyon, except in a few places where it spilled over onto elevated plateaus marginal to the Snake (e.g., Twin Falls). At Farewell Bend the Bonneville Flood flowed through Hell's Canyon and was confined to the Snake River Canyon all the way to Pasco where it joined the Columbia River. The Bonneville flood features identified herein are highlighted in red. SLC = Salt Lake City, Pt = Pocatello, TF = Twin Falls, Bo = Boise, Le = Lewiston, Pa = Pasco, Po = Portland

© Springer Nature Switzerland AG 2021
B. N. Bjornstad, *Ice Age Floodscapes of the Pacific Northwest*, https://doi.org/10.1007/978-3-030-53043-3_5

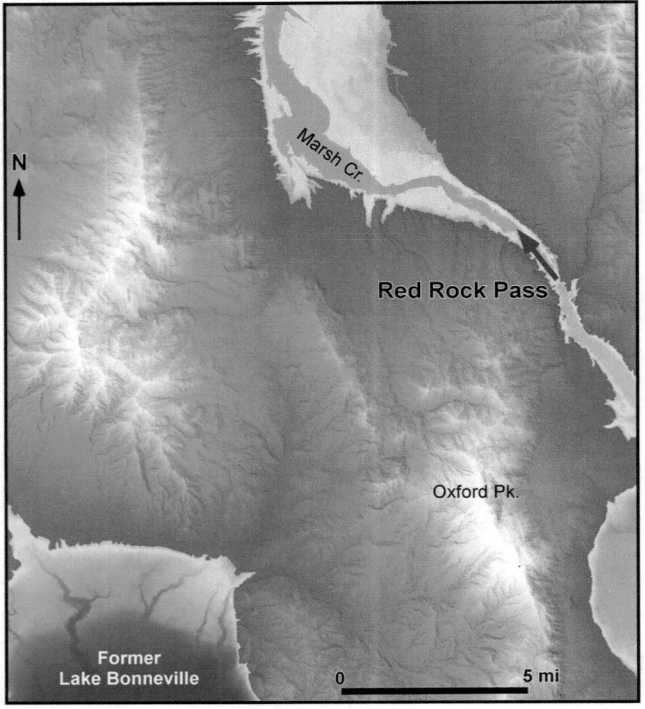

Bonneville Flood Through Red Rock Pass About 18,000 years ago, at the peak of the last Ice Age, a single flood from Utah's pluvial Lake Bonneville occurred as the lake filled to maximum. Floodwater began to escape from the enclosed basin here at the spillover point for the lake at Red Rock Pass. At first, the lake overflowed slowly as it eroded into loose sediment of an alluvial fan complex shed off the surrounding mountains. Once started, however, erosional acceleration of the lake waters rapidly cut a channel through the alluvial dam. The ensuing flood lasted for a couple weeks until floodwaters cut a 400 ft deep channel across the pass. Then the flood probably shut down suddenly upon encountering the resistant bedrock below. Altogether half the volume (~1270 cubic miles) of Lake Bonneville, flowing at up to 57 million cubic ft/sec, escaped through the pass (O'Connor et al., 2020). Image above looks north through Red Rock Pass. Downstream, some flood features very similar to those of the Channeled Scabland lie along Idaho's Snake River valley.

Lake Bonneville Spillover Looking south across Red Rock Pass into the former Lake Bonneville Basin. Drainage began here at the extreme north end of the lake. Spillover channel lies where the pass is narrowest and the incised margins of 400-ft tall alluvial fan are exposed along the channel margins (white arrows).

Massacre Rocks Looking downstream, several horseshoe-shaped recessional cataracts (arrows) eroded away from the main channel of the Snake River. Upper cataract was breached when it eroded through the head of the cataract next to the Snake River. Located 85 flood miles downstream of Red Rock Pass.

Cauldron Linn Intensely eroded basalt was mainly confined to the Snake River canyon but some floodwater locally spilled over onto basalt flatlands above, stripping away all the surficial sediment down to bare basalt bedrock. Notice, however, that soil cover and productive farmlands exist away from area of flood scour. Located 150 flood miles from Red Rock Pass.

Devil's Corral Several recessional cataract canyons (arrows) eroded up a side channel (Eden) before cascading into the Snake River canyon near Twin Falls, Idaho. During this time floodwaters simultaneously filled the Snake River canyon at lower left. Located 160 flood miles from Red Rock Pass.

Melon Gravels along Snake River Boulder Bar Flood-gravel bar at the Birds of Prey National Conservation Area along the Snake River, 250 miles downstream of Red Rock Pass. All boulders are basalt. Most are rounded (like giant watermelons – hence the name Melon Gravel) from flood transport during a single flood from Lake Bonneville. The boulders preferentially collected here at the inside of curve in the Snake River canyon. Some large boulders are angular, indicating they more recently rolled down off adjacent canyon walls. Looking downstream.

Farewell Bend   At left, the Snake River and formerly the Bonneville floodwaters entered 150-mile long Hells Canyon (left) before exiting near Lewiston, ID. This constriction at the head of Hells Canyon caused Bonneville floodwaters to back up temporarily (see page 159). Farewell Bend is located 350 miles downstream of Red Rock Pass. Looking east.

Tammany Bar Outburst-Flood Deposits   Here, within a borrow pit at Lewiston, Idaho is evidence for both the Bonneville Flood, coming down the Snake River, followed by multiple floods from Glacial Lake Missoula that backflooded UP the Snake River from the Columbia Basin. The gravelly, foreset-bedded gray deposits below are from the Bonneville Flood, whereas the overlying pale-brown layers, mostly composed of silt and sand, were deposited during multiple Lake Missoula floods that backflooded UP the Snake River. Dashed line is the contact between Bonneville and Missoula flood deposits. White arrows appear to be ash layers deposited during multiple volcanic eruptions that occurred between a few of the Missoula floods. Located ~500 miles downstream of Red Rock Pass. Looking east.

Glossary

alcove a deep, horseshoe-shaped inner canyon that forms below a recessional cataract.

anastomosis a braided, interlacing network of branching and reuniting flood channels.

anticline a fold in the Earth's crust, arched upward, whose core contains older rocks than its sides.

argillite a kind of metamorphic rock, originally a mudstone or shale, that has since been transformed into an extremely hard, compact, rock tightly cemented with silica. Locally, argillite and associated quartzite come from a group of rocks known as the Belt Supergroup (1.4 billion to 1.6 billion years old) that crops out in northern Washington, Idaho and western Montana.

basalt a dark, fine-grained, volcanic-igneous rock composed primarily of two minerals, plagioclase and pyroxene. Between about 6 million and 17 million years ago, hundreds of lava flows of Columbia River basalt flowed out of long, linear vents in southeastern Washington, eastern Oregon and west-central Idaho and traveled hundreds of miles before cooling and solidifying to form the Columbia Plateau.

Belt rock 1.4 billion- to 1.6 billion-year-old rocks of the Belt Supergroup. Mostly composed of metamorphosed ancient marine sandstone (quartzite) and mudstone (argillite). They are so old they lack fossils of any complex life forms. Belt rocks are a mind-boggling 1000 times older than the Columbia River basalt they underlie.

bioturbation the burrowing, churning and stirring of loose sediment by plants and/or animals near the land surface. Bioturbation is often observed in and associated with paleosols.

borrow pit man-made excavation to remove loose sand and gravel materials for use in roads, construction, etc.

breakout area the area along and immediately in front of the ice dam for glacial Lake Missoula.

butte an isolated, often flat-topped landform that is taller than it is wide. The sides are steep and the lower slopes are often covered with talus. Ice Age floods eroded flat-lying basalt flows to form isolated buttes. The more flood-resistant basalt entablature usually caps the top of buttes.

butte-and-basin topography a general term coined by J Harlen Bretz to describe all landforms created by extreme flood erosion, including rock basins, benches, buttes and mesas within the Channeled Scabland.

caliche a secondary chemical deposit of calcium carbonate that binds sediment or soil together into a rock-like mass. Caliche, which slowly develops in arid to semiarid climates, may take many thousands of years to form.

cataract a waterfall, especially one of great volume in which the vertical descent has been concentrated in one sheer drop over a precipice. Used in areas of Ice Age floods to describe a tall, now-dry cliff formed during cataract recession.

cataract canyon the deep, steep-walled canyon that extends below a recessional cataract.

catastrophism the doctrine that sudden, violent, short-lived events outside our present experience or knowledge of nature have greatly modified the surface of the Earth. J Harlen Bretz promoted catastrophism, unpopular at the time, to his peers to explain the features he observed in the Channeled Scabland. Opposite of uniformitarianism (or gradualism), which was the popular theory at the time. Both coChanneled Scabland: a region of interconnected flood channels, coulees and cataracts eroded through Palouse loess and into basalt by cataclysmic floods in eastern Washington.

Cheney-Palouse Scabland Tract a network of interconnected flood channels and coulees, separated by uplands of Palouse loess, that trends south across the eastern portion of the Channeled Scabland. ncepts have validity.

clay extremely small sedimentary particles that are less than 0.004 millimeters in diameter.

coarse-grained pertains to sedimentary material that is composed of relatively large particles of sand and/or gravel.

colonnade columnar basalt that forms in the lower, interior cooling zone of a basalt lava flow that contains larger, more massive columns, most of which are bounded by vertically oriented cooling fractures. Polygonal columns up to several feet in diameter are common to some flows.

© Springer Nature Switzerland AG 2021
B. N. Bjornstad, *Ice Age Floodscapes of the Pacific Northwest*, https://doi.org/10.1007/978-3-030-53043-3

Opposite of entablature that forms in the upper cooling zone of a basalt flow.

Columbia River basalt   general name given to Miocene age (6 million to 17 million years ago) basalt lava flows of the Columbia Plateau.

Columbia River Lobe   lobe of glacial ice from the Cordilleran Ice Sheet that sometimes extended down the Columbia River Valley from the north. Glacial Lake Spokane would form behind the Columbia River Lobe when the ice dam blocked the Spokane River.

Cordilleran Ice Sheet   name for the ice sheet, thousands of feet thick, that extended across western North America from the Canadian Arctic periodically during the Pleistocene Epoch. Restricted to an area west of the continental divide.

coulee   a long, deep, steep-walled, Ice Age flood channel, now completely dry or occupied by an underfit stream. French for "flow or rush of a torrent." Recessional cataracts may lie along, or at the head of, the coulee.

Coulee Monocline   a long flexure in the surface of the Columbia River basalt that trends along and across Grand Coulee. Along the monocline a narrow band of basalt rock was tilted uniformly to the east and south by tectonic stresses applied within the Earth's crust.

cross stratification   where the internal layering within a bed is inclined at an angle with respect to normal horizontal bedding above and below.

dike   a near-vertical, tabular igneous intrusion that cuts across the bedding or layering in the rock that is intruded.

divide crossing   where deep floodwater flowed over a low divide separating two drainages, often eroding a notch or spillover channel across the abandoned spillway.

eddy bar   a flood bar that forms where eddy currents develop along a flood route. Often found in the more protected mouths of tributary valleys where they join main flood pathways.

entablature   the upper portion of basalt flows that show a more dense and random distribution of cooling fractures. In contrast to colonnade that forms at or toward the bottom of flows.

erratic   an out-of-place rock (often boulder size) that is of a different rock type than the bedrock beneath it and transported to its present location by glaciers or floating icebergs. Edges of erratics are typically, but not always, angular.

erratic cluster   a grouping of ice-rafted erratics that probably melted out from the same iceberg after becoming grounded.

expansion bar   a giant flood bar that forms where flood channels or coulees suddenly widen or expand. As the water expands it also slows down from the venturi effect, which results in sediment buildup and bar formation.

faceted escarpment   an over-steepened slope or cliff planed off by the movement of strong currents along a flood channel. Often a series of individual escarpments may align along channel margins.

fine-grained   pertains to sedimentary material in flood deposits composed of mostly small particles of silt and/or fine sand.

flood bar   an accumulation of sediment, most often composed of sand and/or gravel, that occurs along flood channels where the currents temporarily slowed for various reasons. Different types of flood bars from the Ice Age floods in the Channeled Scabland include eddy, expansion, longitudinal, crescent and pendant bars.

flood channel   straight to curved paths scoured out by the floods. Deep, box-shaped flood channels are called coulees.

foreset bedding   internal bedding within a sedimentary deposit that is inclined to the principal surface of accumulation. Foreset beds generally dip in the direction of water flow.

fracture   a general term for any naturally broken surface in rock. In the Channeled Scabland, small fractures developed during cooling and contraction of basalt lava. Later, long after the lava hardened, more extensive parallel sets of fractures occasionally developed locally from larger, regional tectonic stresses within the crust.

giant current ripples   waves of sand and gravel deposited by deep, fast moving cataclysmic flood currents, similar to ripples along a river or on the beach, but much, much larger.

gneiss   a light- and dark-banded metamorphic rock composed predominantly of quartz, feldspar and hornblende.

grading   refers to a progressive shift in the size of sedimentary particles within a bed. Normal grading is a shift from larger to smaller particles upward within the bed; reverse grading is the opposite.

Grand Coulee   a 50-mile-long coulee within the western Channeled Scabland that displays some of the most recent and intense erosional flood features anywhere on Earth.

granite   a light-colored, crystalline, intrusive-igneous rock that contains abundant minerals of gray to white quartz and feldspar, in addition to pink-colored alkali feldspar.

granitic   a general term for any light-colored intrusive-igneous rock that formed deep underground from a cooling body of liquid magma. Granitic rocks include granite and granodiorite. Most ice-rafted erratics in the Channeled Scabland are granitic in origin; granitic rocks underlie the head of Grand Coulee and also occur north and east of the Channeled Scabland.

granodiorite   a light-colored, crystalline, intrusive-igneous rock that contains abundant minerals of gray to white quartz and feldspar.

gravel sedimentary particles larger than sand (more than 2 millimeters in diameter). Gravel clasts are referred to, in increasing size, as granules, pebbles, cobbles and boulders.

grooves in the Columbia Plateau, refers to a series of long, roughly parallel ridges or flutings on the land surface due to intense flood erosion. Straight to curvilinear grooves are often visible on flood-swept upland plateaus.

hanging coulee a side coulee that enters a larger coulee at a higher elevation. Characteristic of the Channeled Scabland, where flat valley floors suddenly drop off into nothingness at either one or both ends.

Holocene an epoch of geologic time between the present and the end of the last Ice Age (about 13,000 years ago).

hydraulic constriction places along the flood route where a large volume of water was forced to go through a narrow opening. If more water enters the opening than can drain through, then the constriction will cause water to back up, creating a type of hydraulic dam (e.g., Wallula Gap). Because of the venturi effect, water naturally speeds up through hydraulic constrictions.

ice lobe a finger of ice that extends south from an ice sheet. Two ice lobes of the Cordilleran Ice Sheet, the Purcell Trench Lobe and Okanogan Lobe, had a direct impact on the timing and frequency of Ice Age floods in the Pacific Northwest.

ice dam formed where an ice lobe moved across and blocked normal stream drainage. Glacial lakes, up to thousands of feet deep, filled behind ice dams.

igneous rock that solidified from molten or partly molten material, including both intrusive and extrusive rock types. One of the three principal rock types, along with sedimentary and metamorphic.

inner canyon steep-walled, deep canyon or coulee below a recessional cataract or incised into a larger canyon.

Lake Columbia a glacial lake that formed behind the Okanogan Lobe of glacial ice.

Lake Spokane a glacial lake that formed behind the Columbia River Lobe of glacial ice, near the junction of the Columbia and Spokane rivers.

Lake Missoula the glacial lake, up to 2000 feet deep, responsible for most, but not all, of the Ice Age floods. The lake formed as the Purcell Trench Lobe blocked the Clark Fork River, creating an ice dam up to 3000 feet high in the Idaho Panhandle. The lake failed episodically, sending up to 500 cubic miles of water downstream. At its maximum, Lake Missoula was 200 miles long and covered a 3000-square-mile area. It took up to 60 years for glacial meltwater to refill the lake but only a few days to empty.

landslide the downslope movement of weakened soil or rock en masse under the influence of gravity.

loess windblown silt and fine sand. Among flood routes it has collected downwind of sedimentary basins, especially in the Palouse country where it occurs as rolling hills up to 250 feet thick. Loess deposition began forming about the same time as the earliest Ice Age floods (about 2 million years ago) and continues to form today.

magnetic polarity reversal refers to a flip in the Earth's magnetic field that has occurred periodically through geologic time. The last magnetic-polarity shift that caused a reversal in the magnetic field occurred 780,000 years ago.

mesa an isolated, nearly level rock mass standing distinctly above the surrounding countryside; mesas are bounded by abrupt, steep-sided slopes on all sides and capped by erosion-resistant rock such as entablature basalt. Width of a mesa is always much greater than its height, which distinguishes it from a butte.

metamorphic a rock derived from pre-existing, deeply buried rocks and altered due to intense heat and/or pressure. One of the three principal rock types, along with sedimentary and igneous.

metasedimentary sedimentary rocks that show both sedimentary and metamorphic characteristics. Most rocks of the 1.5 billion-year-old Belt Supergroup fit into this category.

Miocene an epoch of geologic time spanning an interval from 5 million to 24 million years ago. The Columbia River basalt all poured out of vents during the Miocene Epoch between about 6 million and 17 million years ago.

Missoula flood an outburst flood derived from the sudden release of ice-dammed glacial Lake Missoula, after the lower end of the Purcell Trench Lobe of ice failed.

monolith a large, upstanding mass of rock. Within the Channeled Scabland, Grand Coulee's Steamboat Rock is the best example.

moraine an accumulation of glacial till with a distinct hummocky form carried and deposited directly by glacial ice.

Moses Coulee a 50-mile-long coulee in the western Channeled Scabland used during some earlier floods but blocked from most later floods by the Okanogan Lobe.

Okanogan Lobe a lobe of the Cordilleran Ice Sheet that extended south from Canada along the Okanogan River Valley and blocked the Columbia River near present-day Grand Coulee Dam. Glacial Lake Columbia was the ice-dammed lake that backed up behind the Okanogan Lobe.

outburst plain used here to describe the broad plain of outburst-flood sediments deposited immediately downstream of the breakout area for the Missoula floods along Rathdrum Prairie.

paleochannel a remnant of a channel cut into older rock or sediments that later became filled with younger sediments.

paleomagnetic pertains to the natural magnetization that resides in rock and sediment material and records the intensity and direction of the Earth's magnetic field when the material was emplaced.

paleosol a buried soil horizon of the geologic past.

Palouse name given the treeless rolling hills of up to 250-foot-thick windblown loess of eastern Washington. Derived from the French word "pelouse" meaning "lawn" or "grassy plains." Palouse soils originally supported native bunch grass but today are used to grow mostly dryland wheat.

Palouse Slope a gentle, southwest-sloping surface that underlies the Palouse region. The surface is a reflection of the top of the Columbia River basalt, which has a similar attitude. The Palouse Slope merges with the Yakima Folds to the west.

particle a discrete, individual sedimentary grain or fragment.

patterned ground arrangements of surface materials produced by frost action or other natural processes into various geometric forms.

pendant bar a type of flood bar that forms immediately downstream of a flow obstruction within a flood channel or coulee.

pillar used here to describe a tall, narrow eroded basalt monolith with an irregular top. Height of a pillar is much greater than its width.

pillow basalt pillows appear as large, rounded balls of glassy basalt at the base of the lava flow. The pillows indicate the lava cooled very suddenly after coming in contact with a lake or stream.

Pleistocene an epoch of geologic time between about 13,000 years and 2.6 million years ago. The Pleistocene essentially spans the same time period known as the Ice Age.

plexus a complex structure containing an intricate network of parts.

plunge pool a deep, circular hole eroded at the base of a cataract, scoured out by the force of up to hundreds of feet of floodwater rushing over the cataract.

pluvial having to do with rain or much rain.

pothole a circular rock basin eroded into bedrock, which was basalt in the case of the Channeled Scabland. Many are the result of circulating, tight eddies or kolk-like currents during flooding.

Precambrian All geologic time prior to 570 million years ago. Comprises 90 percent of all geologic time on Earth.

Purcell Trench Lobe A lobe of ice that came down the Purcell Trench from Canada, blocking the Clark Fork River along the present-day Idaho-Montana border. Glacial Lake Missoula, responsible for most of the Ice Age floods, formed behind this ice dam.

quartzite an extremely hard metamorphic rock composed entirely of quartz. Derived from an ancient sandstone, quartzite may display some original horizontal or cross stratification that developed as the sandstone was deposited. Quartzite and associated argillite clasts found in the Columbia Plateau are mostly from Belt rocks (1.4 billion to 1.6 billion years old) present in northern Washington, Idaho and Montana. The quartzite from these regions is frequently pale blue, green or brown.

recessional cataract a tall, steep cliff that forms from the upstream migration of a cataract in a flood channel or coulee during the Ice Age floods. A horseshoe-shaped alcove commonly lies at the head of the cataract while a deep inner canyon lies below the cataract.

rhythmite a graded sedimentary bed, several inches to several feet thick, deposited under slackwater conditions, especially in backflooded valleys during Ice Age flooding. Some believe that each rhythmite represents a separate outburst flood from glacial Lake Missoula.

ringed crater unusual circular features in Columbia River basalt etched out by the Ice Age floods. Concentric rings are basalt dikes that formed as the basalt lava came to rest and cooled.

rip-up clast large fragments of ripped up, sedimentary material eroded and transported from Palouse loess, or older flood deposits during Ice Age flooding. Because rip-up clasts are extremely friable and fall apart in their present state, they were probably frozen before being ripped up and transported by floods.

rock basin In the Channeled Scabland a depression carved out in weaker basalt bedrock during flooding.

rock bench a tier of basalt, eroded by the floods, that lies alongside or above a flood channel or coulee. Rock benches usually conform to the tops of basalt lava flows.

rock blade a narrow and tall ridge or rib of basalt within a flood coulee, often separating adjacent coulees, associated with cataract recession.

rock shelter a cave-like opening that extends a short distance into a hill or cliff side. Oftentimes associated with erosional scouring of weaker rock between or within basalt flows.

sand small sedimentary particles that are between 0.06 millimeters to 2 millimeters in diameter.

scabland flood Ice Age flood that flowed across the Channeled Scabland of southeastern Washington. Most scabland floods were from glacial Lake Missoula during a time when the Okanogan Lobe obstructed flow and diverted the floodwaters out across the scabland.

scabland tract a broad area with multiple interconnected flood channels eroded through Palouse loess down to basalt bedrock.

scarp a planar, beveled slope. In the Channeled Scabland scarps were eroded in bedrock and in the Palouse

Formation during erosion along flood channels. Scarp slopes are steeper than adjacent non-flood-eroded slopes.

sediment   solid fragmental material that originates from the weathering and transport, by water, wind or ice, of pre-existing rock or sediment bodies.

sedimentary   pertains to an accumulation of sediment particles, usually under the influence of moving water or wind. One of the three principal rock types, along with igneous and metamorphic.

sequence   a succession of geologic events recorded in beds of rock and/or sediment. Lower beds record events that happened earlier than beds higher in the sequence.

silt   tiny sedimentary particles that are 0.004 millimeters to 0.06 millimeters in diameter.

slackwater   refers to areas with slower moving floodwaters associated with Ice Age flooding (such as backflooded valleys and valley margins) where fine-grained sediment (mostly sand and silt) was deposited.

spillover channel   a divide crossing where floodwaters erode enough to carve a somewhat flat-bottomed, trapezoidal-shaped channel across the divide.

steptoe   a hill or ridge of older rock surrounded by engulfing basalt lava flows.

strandline   like a bathtub ring, the level at which a former standing body of water met the land surface. Strandlines may be visible as small terraces etched out by wave activity along former lake shorelines.

talus   large, angular rock fragments derived from and lying at the base of cliffs. Most talus on the Channeled Scabland is composed of basalt fragments shed off steep cliffs since the last Ice Age floods. Also known as scree.

tectonic   pertains to large, regional forces deep within the Earth's crust that cause earthquakes, folds, faults and some fractures observed at or near the surface.

Telford-Crab Creek Scabland Tract   a network of interconnected flood channels and coulees, separated by uplands of Palouse loess, that trends south across the central portion of the Channeled Scabland.

till   an unsorted, heterogeneous mixture of clay, silt, sand and gravel deposited beneath a slowly advancing glacier.

trenched spur   in flood coulees, where the depth, force and momentum of the floodwaters forced the flow to go straight over localized basalt uplands (spurs), cutting new channels across the spurs.

trimline   a distinct boundary along flood channels that separates a lower, eroded area, without soil cover, from a higher area with soil cover preserved. The trimline generally marks the upper height limit for the floods.

unconformity   refers to the contact between beds where one layer is significantly different in age than the over-lying bed. Signifies an erosional event or period of non-deposition between beds.

stream underfit   a stream that appears to be too small to have eroded the valley in which it flows.

uniformitarianism   The fundamental geologic principle or doctrine that geologic processes and natural laws now operating to modify the surface of Earth have acted in the same regular manner and with the same intensity throughout geologic time and that past geologic events can be explained by phenomenon and forces observable today. Also referred to as gradualism. Opposite of catastrophism.

upland plateau   an elevated, broad basalt plateau or mesa swept clean of Palouse loess and other soil by high-energy Ice Age floods. The plateau surface is usually riddled with erosional longitudinal grooves, potholes and rock basins. Upland plateaus are bordered by deeper erosional flood channels or coulees.

varve   an annual layer of sediment deposited within a glacial lake. Typically, lighter-colored fine sand to silt at the base of a varve transitions upward into darker clay, representing summer versus winter layers, respectively. Normally many varves, each only fractions of an inch thick, occur stacked together within a sequence.

venturi effect   the physical phenomenon that water forced to move through a narrower channel will move faster than the same amount of water moving through a wider channel.

vesicle   a cavity of variable shape and size in a lava flow, formed by the entrapment of a gas bubble as the lava cooled and solidified.

Waterville Plateau   a broad, gently rolling upland that lies between Grand Coulee and the Columbia River valley within the western Channeled Scabland. Much of the Waterville Plateau was covered by the Okanogan Lobe and thus not affected by Ice Age floods, except through Moses Coulee.

weathering   alteration of sediment or rock at the Earth's surface by long-term exposure to air, water and other agents through various physical, chemical and biological processes.

Withrow Moraine   a well-defined ridge of glacial debris (mostly glacial till and boulders) deposited at the farthest extent of the Okanogan Lobe on the Waterville Plateau.

Yakima Folds   Roughly parallel anticlines of folded Columbia River basalt in portions of southern and central Washington. Combined, the Yakima Folds comprise a geologic region called the Yakima Fold Belt.

Bibliography

Allen, J. E., Burns, M., & Burns, S. (2009). *Cataclysms on the Columbia: The great Missoula floods* (2nd ed.). Portland: Ooligan Press. 204p.

Allison, I. S. (1935). Glacial erratics in Willamette valley. *Geological Society of America Bulletin, 46,* 615–632.

Allison, I. S. (1978). Late Pleistocene sediments and floods in the Willamette valley. *The Ore Bin, 40*(11–12), 177–202.

Alt, D. D. (2001). *Glacial Lake Missoula and its humongous floods.* Missoula: Mountain Press. 199p.

Atwater, B. F. (1984). Periodic floods from glacial Lake Missoula into the Sanpoil arm of glacial Lake Columbia. *Geology, 12,* 464–467.

Atwater, B. F. (1986). *Pleistocene glacial-lake deposits of the Sanpoil River valley, northeastern Washington.* U.S. Geological Survey Bulletin 1661, 39p.

Atwater, B. F. (1987). Status of glacial Lake Columbia during the last floods from glacial Lake Missoula. *Quaternary Research, 27,* 182–201.

Baker, V. R. (1973). *Paleohydrology and sedimentology of Lake Missoula flooding in eastern Washington.* GSA Special Paper No. 144, Boulder: Geological Society of America, 73p.

Baker, V. R. (Ed.). (1981). *Catastrophic flooding: The origin of the Channeled Scabland, benchmark papers in geology.* Stroudsburg/ Pennsylvania: Dowden, Hutchinson & Ross. 360p.

Baker, V. R., & Bunker, R. C. (1985). Cataclysmic late Pleistocene flooding from glacial Lake Missoula: A review. *Quaternary Science Reviews, 4,* 1–41.

Baker, V. R. (1987). Dry Falls of the Channeled Scabland, Washington. In M. L. Hill (Ed.), *Centennial field guide* (Vol. 1, pp. 369–372). Boulder: Cordilleran Section of the Geological Society of America.

Baker, V. R. (1989). The Grand Coulee and Dry Falls. In R. M. Breckenridge (Ed.), *Glacial Lake Missoula and the Channeled Scabland* (pp. 51–57). American Geophysical Union, Field Trip Guidebook T310, Chapter 6.

Baker, V. R. (1995). Surprise endings to catastrophism and controversy on the Columbia: Joseph Thomas Pardee and the Spokane flood controversy. *GSA Today, 5,* 169–173.

Baker, V. R. (2002). The study of superfloods. *Science, 295,* 2379–2380.

Baker, V. R. (2009a). Channeled Scabland morphology. In D. M. Burr, P. A. Carling, & V. R. Baker (Eds.), *Megaflooding on Earth and Mars* (pp. 65–77). Cambridge: Cambridge University Press.

Baker, V. R. (2009b). The Channeled Scabland: A retrospective. *Annual Review of Earth and Planetary Science, 37,* 393–411.

Baker, V. R., & Nummedal, D. (Eds.). (1978). *The Channeled Scabland.* Washington, D.C.: National Aeronautics and Space Administration. 186p.

Baker, V. R., Bjornstad, B. N., Busacca, A. J., Fecht, K. R., Kiver, E. P., Moody, U. L., Rigby, J. G., Stradling, D. F., & Tallman, A. M. (1991). *Quaternary geology of the Columbia Plateau.* In R. B. Morrison (Ed.), *Quaternary nonglacial geology; conterminous U.S., the geology of North America* (Vol. K-2, pp. 215–250). Boulder: Geological Society of America.

Balbas, A. M., Barth, A. M., Clark, P. U., Clark, J., Caffee, M., O'Connor, J. E., Baker, V. R., Konrad, K., & Bjornstad, B. N. (2017). 10Be dating of late Pleistocene megafloods and Cordilleran Ice Sheet retreat in the northwestern United States. *Geology, 45,* 583–586.

Benito, G., & O'Connor, J. E. (2003). Number and size of last-glacial Missoula floods in the Columbia River valley between Pasco Basin, Washington and Portland, Oregon. *Geological Society of America Bulletin, 115,* 624–638.

Bjornstad, B. N. (1980). *Sedimentology and depositional environment of the Touchet Beds, Walla Walla basin, Washington.* Master's Thesis. Cheney: Eastern Washington University, 138p.

Bjornstad, B. N., Fecht, K. R., & Pluhar, C. J. (2001). Long history of pre-Wisconsin, Ice Age floods: Evidence from southeastern Washington state. *Journal of Geology, 109,* 695–713.

Bjornstad, B. N. (2006). *On the trail of the Ice Age floods: A geological field guide to the mid-Columbia basin.* Sandpoint: Keokee Publishing. 308p.

Bjornstad, B. N., Babcock, R. S., & Last, G. V. (2007). Flood basalts and Ice Age floods: Repeated cataclysms of southeastern Washington. In P. Stelling & D. S. Tucker (Eds.), *Floods, faults, and fire: Geological field trips in Washington State and Southwest British Columbia* (Vol. 9, pp. 209–255). Boulder: Geological Society of America Field Guide. https://doi.org/10.1130/2007/fld009(10).

Bjornstad, B. N., & Kiver, E. P. (2012). *On the trail of the Ice Age floods: The northern reaches.* Sandpoint: Keokee Publishing. 432p.

Bjornstad, B. N. (2014). Ice-rafted erratics and bergmounds from Pleistocene outburst floods, Rattlesnake Mountain, Washington, USA. *Quaternary Science Journal, 63,* 44–59. https://doi.org/10.3285/eg.63.1.03.

Booth, D. B., Troost, K. G., Clague, J. J., & Waitt, R. B. (2004). The Cordilleran Ice Sheet. In A. R. Gillespie et al. (Eds.), *The Quaternary Period in the United States: Developments in Quaternary science* (Vol. 1, pp. 17–44). Amsterdam: Elsevier.

Breckenridge, R. M. (1989a). Evidence for ice dams and floods in the Purcell Trench: Trip A. In N. L. Joseph (Ed.), *Geologic guidebook for Washington and adjacent areas* (Information Circular 86) (pp. 309–320). Olympia: Washington Division of the Geology and Earth Resources.

Breckenridge, R. M. (1989b). Lower glacial Lakes Missoula and Clark Fork ice dams. In R. M. Breckenridge (Ed.), *Glacial Lake Missoula and the Channeled Scabland* (pp. 13–22). American Geophysical Union, Field Trip Guidebook T310, Chapter 3.

Breckenridge, R. M., & Sprenke, K. F. (1997). An overdeepened glaciated basin, Lake Pend Oreille, northern Idaho. *Glacial Geology and Geomorphology, 3,* 1–10.

Bretz, J. H. (1923). The Channeled Scabland of the Columbia Plateau. *Journal of Geology,* 617–649.

Bretz, J. H. (1925). The Spokane flood beyond the Channeled Scablands. *Journal of Geology, 33,* 97–259.

Bretz, J. H. (1927). Channeled Scabland and the Spokane flood. *Journal of Washington Academy of Sciences, 18,* 200–211.

© Springer Nature Switzerland AG 2021
B. N. Bjornstad, *Ice Age Floodscapes of the Pacific Northwest*, https://doi.org/10.1007/978-3-030-53043-3

Bretz, J. H. (1928a). The Channeled Scabland of eastern Washington. *Geographical Review, 18*, 446–477.

Bretz, J. H. (1928b). Bars of Channeled Scabland. *Geological Society of America Bulletin, 39*, 643–702.

Bretz, J. H. (1928c). Alternate hypotheses for Channeled Scabland. *Journal of Geology, 36*, 193–223, 312–193–223, 341.

Bretz, J. H. (1930). Lake Missoula and the Spokane flood. *Geological Society of America Bulletin, 41*, 92–93.

Bretz, J. H. (1932). *The Grand Coulee.* American Geographical Society, Special Publication No. 15, New York, 89p.

Bretz, J. H. (1959). *Washington's Channeled Scabland.* Bulletin No. 45, Washington Division of Mines and Geology, Olympia.

Bretz, J. H. (1969). The Lake Missoula floods and the Channeled Scabland. *Journal of Geology, 77*, 505–543.

Bretz, J. H., Smith, H. T. U., & Neff, G. E. (1956). Channeled Scabland of Washington: New data and interpretations. *Geological Society of America Bulletin, 67*, 957–1049.

Brunner, C. A., Normark, W. R., Zuffa, G. G., & Serra, F. (1999). Deep-sea sedimentary record of the late Wisconsin cataclysmic floods from the Columbia River. *Geology, 27*, 463–466.

Burns, W. J., & Coe, D. E. (2012). *Missoula floods—Inundation extent and primary flood features in the Portland metropolitan area, Clark, Cowlitz, and Skamania Counties, Washington, and Clackamas, Columbia, Marion, Multnomah, Washington, and Yamhill Counties, Oregon,* IMS-36. Oregon: Department of Geology and Mineral Industries.

Burr, D. M., Carling, P. A., & Baker, V. R. (Eds.). (2019). *Megaflooding on Earth and Mars* (pp. 172–193). Cambridge: Cambridge University Press.

Busacca, A. J., & McDonald, E. V. (1994). Regional sedimentation of late Quaternary loess on the Columbia Plateau: Sediment source areas and loess distribution patterns. *Washington Division of Geology and Earth Resources Bulletin, 80*, 181–190.

Busacca, A. J., McDonald, E. V., & Baker, V. R. (1989). The record of pre-late Wisconsin floods and late Wisconsin flood features in the Cheney-Palouse scabland. In R. M. Breckenridge (Ed.), *Glacial Lake Missoula and the Channeled Scabland* (pp. 57–62). American Geophysical Union, Field Trip Guidebook T310, Chapter 7.

Busacca, A. J. (1989). Long Quaternary record in eastern Washington, U.S.A., interpreted from multiple buried paleosols in loess. *Geoderma, 45*, 105–122.

Carrara, P. E., Kiver, E. P., & Stradling, D. F. (1996). The southern limit of Cordilleran ice in the Colville and Pend Oreille valleys of northeastern Washington during the late Wisconsin glaciations. *Canadian Journal of Earth Sciences, 33*, 769–778.

Carson, R. J., & Pogue, K. R. (1996). *Flood basalts and glacier floods: Roadside geology of parts of Walla Walla, Franklin, and Columbia Counties, Washington,* Washington Division of Geology and Earth Resources. Information Circular, 90, 47p.

Clarke, G. K. C., Matthews, W. H., & Pack, R. T. (1984). Outburst floods from glacial Lake Missoula. *Quaternary Research, 22*, 289–299.

Crosby, C. J., & Carson, R. J. (1999). Geology of Steamboat Rock, Grand Coulee, Washington. *Washington Geology, 27*, 3–8.

Denlinger, R. P., & O'Connell, D. R. H. (2010). Simulations of cataclysmic outburst floods from Pleistocene glacial Lake Missoula. *Geological Society of America Bulletin, 122*, 678–689. https://doi.org/10.1130/B26454.1.

Fecht, K. R., Reidel, S. P., & Tallman, A. M. (1987). Paleodrainage of the Columbia River on the Columbia Plateau of Washington State – A summary. In *Bulletin 77, Washington Division of Geology and Earth Resources* (pp. 219–248). Olympia: Department of Natural Resources.

Gaylord, D. R., Busacca, A. J., & Sweeney, M. R. (2003). The Palouse loess and the Channeled Scabland: A paired Ice Age geologic system. In D. J. Easterbrook (Ed.), *Quaternary geology of the United States, INQUA 2003 field guide volume* (pp. 123–134). Reno: Desert Research Institute.

Hanson, L. G. (1970). *The origin and development of Moses Coulee and other scabland features on the Waterville Plateau, Washington.* PhD Dissertation. Seattle: University of Washington, 137p.

Hanson, M. A., Lian, O. B., & Clague, J. J. (2011). The sequence and timing of large Pleistocene floods from glacial Lake Missoula. *Quaternary Science Reviews, 31*, 67–81.

Hanson, M. A., & Clague, J. J. (2016). Record of glacial Lake Missoula floods in glacial Lake Columbia. *Quaternary Science Reviews, 133*, 62–76.

Hendy, I. L. (2009). A fresh perspective on the Cordilleran Ice Sheet. *Geology, 37*, 95–96.

Karlson, R. C. (2006). *Investigations of the Ice Age flood geomorphology and stratigraphy in Ginkgo Petrified Forest State Park, Washington: Implications for park interpretation.* Master's Thesis. Ellensburg: Central Washington University, 146p.

Kasbohm, J., & Schoene, B. (2018). Rapid eruption of the Columbia River flood basalt and correlation with mid-Miocene climate optimum. *Science Advances, 4*(9), 8.

Keszthelyi, L. P., Baker, V. R., Jaeger, W. L., Gaylord, D. R., Bjornstad, B. N., Greenbaum, N., Self, S., Thordarson, T., Porat, N., & Zreda, M. G. (2009). Floods of water and lava in the Columbia River basin: Analogs for Mars. In J. E. O'Connor, R. J. Dorsey, & I. P. Madin (Eds.), *Volcanoes to vineyards: Geologic field trips through the dynamic landscape of the Pacific Northwest* (Vol. 15, pp. 845–874). Geological Society of America Field Guide. https://doi.org/10.1130/2009.fld015.

Kiver, E. P., & Stradling, D. F. (1989). The Spokane valley and northern Columbia Plateau. In R. M. Breckenridge (Ed.), *Glacial Lake Missoula and the Channeled Scabland* (pp. 23–36). American Geophysical Union, Field Trip Guidebook T310, Chapter 4.

Kiver, E. P., Stradling, D. F., & Moody, U. (1989). Glacial and multiple flood history of the northern borderlands: Trip B. In N. L. Joseph (Ed.), *Geologic guidebook for Washington and adjacent areas* (Information Circular 86) (pp. 321–335). Olympia: Washington Division of the Geology and Earth Resources.

Komar, P. D. (1983). Shapes of streamlined islands on Earth and Mars experiments and analysis of the minimum-drag form. *Geology, 11*, 651–655.

Kovanen, D. J., & Slaymaker, O. (2004). Glacial imprints of the Okanogan Lobe, southern margin of the Cordilleran Ice Sheet. *Journal of Quaternary Science, 19*, 547–565.

Lopes, C., & Mix, A. C. (2009). Pleistocene megafloods in the northeast Pacific. *Geology, 37*, 79–82.

McKee, B., & Stradling, D. F. (1970). The sag flowout: A newly described volcanic structure. *Geological Society of America Bulletin, 81*, 2035–2044.

McDonald, E. V., & Busacca, A. J. (1988). Record of pre-late Wisconsin giant floods in the Channeled Scabland interpreted from loess deposits. *Geology, 16*, 728–731.

McDonald, E. V., & Busacca, A. J. (1989). Record of pre-Late Wisconsin floods and of Late Wisconsin flood features in the Cheney-Palouse Scabland: Trip C. In N. L. Joseph (Ed.), *Geologic guidebook for Washington and adjacent areas, information circular 86* (pp. 337–346). Olympia: Washington Division of the Geology and Earth Resources.

McDonald, E. V., & Busacca, A. J. (1992). Late Quaternary stratigraphy of loess in the Channeled Scabland and Palouse regions of Washington State. *Quaternary Research, 38*, 141–156.

Mueller, M., & Mueller, T. (1997). *Fire, faults and floods.* Moscow: University of Idaho Press. 288p.

Mullineaux, D. R., Wilcox, R. E., Ebaugh, W. F., Fryxell, R., & Rubin, M. (1978). Age of the last major scabland flood of the Columbia Plateau in eastern Washington. *Quaternary Research, 10*, 171–180.

Neff, G. E. (1989). Columbia basin project. In *Vol. 1 of Engineering geology in Washington, Bulletin No. 78* (pp. 535–563). Olympia: Washington Division of Geology and Earth Resources.

Normark, W. R., & Reid, J. A. (2003). Extensive deposits on the Pacific Plate from late Pleistocene North American glacial lake outbursts. *Journal of Geology, 111*, 617–637.

O'Connor, J. E., & Baker, V. R. (1992). Magnitudes and implications of peak discharges from glacial Lake Missoula. *Geological Society of America Bulletin, 104*, 267–279.

O'Connor, J. E. (1993). *Hydrology, hydraulics, and geomorphology of the Bonneville flood*. Geological Society of America Special Paper #274, Geological Society of America, Boulder, Colorado, 83p.

O'Connor, J. E., Baker, V. R., Waitt, R. B., Smith, L. N., Cannon, C. M., George, D. L., & Denlinger, R. P. (2020). The Missoula and Bonneville floods – A review of Ice-Age megafloods in the Columbia River basin. *Earth Science Reviews.* https://doi.org/10.1016/j.earscirev.2020.103181.

Pardee, J. T. (1910). The glacial Lake Missoula. *Journal of Geology, 18*, 376–386.

Pardee, J. T. (1942). Unusual currents in glacial Lake Missoula, Montana. *Geological Society of America Bulletin, 53*, 1569–1599.

Patton, P. C., & Baker, V. R. (1978). New evidence for pre-Wisconsin flooding in the Channeled Scabland and eastern Washington. *Geology, 6*, 567–571.

Reidel, S. P., Camp, V. E., Ross, M. E., Wolff, J. A., Martin, B. S., Tolan, T. L., & Wells, R. E. (2013). *The Columbia River Flood Basalt Province, Geological Society of America Special Paper 497*. Boulder: Geological Society of America. 440p.

Shaw, J., Munro-Stasuik, M., Sawyer, B., Beaney, C., Lesemann, J.-E., Musacchio, A., Rains, B., & Young, R. R. (1999). The Channeled Scabland: Back to Bretz. *Geology, 27*, 605–608.

Smith, G. A. (1993). Missoula flood dynamics and magnitudes inferred from sedimentology of slack-water deposits on the Columbia Plateau. *Geological Society of America Bulletin, 195*, 77–100.

Smyers, N. B., & Breckenridge, R. M. (2003). Glacial Lake Missoula, Clark Fork ice dam, and the floods outburst area: Northern Idaho and western Montana. In T. W. Swanson (Ed.), *Western Cordillera and adjacent areas, Field Guide 4* (pp. 1–15). Boulder: Geological Society of America.

Soennichsen, J. R. (2008). *Bretz's Flood*. Seattle: Sasquatch Books. 289p.

Steele, W. K. (1991). Paleomagnetic evidence for repeated glacial Lake Missoula floods from sediments of the Sanpoil Valley, northeastern Washington. *Quaternary Research, 35*, 197–207.

Sweeney, M. R. (2004). *Sedimentology, paleoclimatology and geomorphology of a late Pleistocene-Holocene paired eolian system, Columbia Plateau*. PhD Dissertation. Pullman: Washington State University, 204p.

Sweeney, M. R., Busacca, A. J., Richardson, C. A., Blinnikov, M., & McDonald, E. C. (2004). Glacial anticyclone recorded in Palouse loess of northwestern United States. *Geology, 32*(8), 705–708.

Thompson, R. (2015). *Gigaflood: The largest of the Lake Missoula floods in northwest Oregon and southwest Washington*. Portland: LMF Publishing. 206p.

Waitt, R. B. (1980). About forty last-glacial Lake Missoula jökulhlaups through southern Washington. *Journal of Geology, 88*, 653–679.

Waitt, R. B., & Thorson, R. M. (1983). The Cordilleran Ice Sheet in Washington, Idaho, and Montana. In H. E. Wright (Ed.), *Late Quaternary Environments of the United States, Volume 1, The Late Pleistocene* (pp. 53–70). Minneapolis: University of Minnesota Press.

Waitt, R. B. (1984). Periodic jokulhlaups from Pleistocene glacial Lake Missoula – New evidence from varved sediment in northern Idaho and Washington. *Quaternary Research, 22*, 46–58.

Waitt, R. B. (1985). Case for periodic, colossal jökulhlaups from Pleistocene glacial Lake Missoula. *Geological Society of America Bulletin, 96*, 1271–1286.

Waitt, R. B. (1987). *Evidence for dozens of stupendous floods from glacial Lake Missoula in eastern Washington, Idaho and Montana*. In M. L. Hill (Ed.), *Centennial field guide 1, trip no. 77* (pp. 345–350). Boulder: Geological Society of America.

Waitt, R. B., & Atwater, B. F. (1989). Stratigraphic and geomorphic evidence for dozens of last-glacial floods. In R. M. Breckenridge (Ed.), *Glacial lake Missoula and the Channeled Scabland* (pp. 37–50). American Geophysical Union, Field Trip Guidebook T310, Chapter 5.

Waitt, R. B. (1994). Scores of gigantic, successively smaller Lake Missoula floods through Channeled Scabland and Columbia valley. In D. A. Swanson & R. A. Haugerud (Eds.), *Geologic field trips in the Pacific Northwest, Chapter 1K* (pp. 1K-1–1K-88). Boulder: Geological Society of America.

Waitt, R. B., Denlinger, R. P., & O'Connor, J. E. (2009). Many monstrous floods down Channeled Scabland and Columbia valley. In J. E. O'Connor, R. J. Dorsey, & I. P. Madin (Eds.), *Volcanoes and vineyards: Geologic field trips through the dynamic landscape of the Pacific Northwest* (Vol. 15, pp. 775–844). Geological Society of America Field Guide.

Waitt, R. B., Long, W. A., & Stanton, K. M. (2019). Erratics and other evidence of late-Wisconsin Missoula outburst floods in lower Wenatchee and adjacent Columbia valleys, Washington. *Northwest Science, 92*, 318–337.

Weis, P. L., & Newman, W. L. (1989). *The Channeled Scablands of eastern Washington: The geologic story of the Spokane flood* (2nd ed.). Cheney: Eastern Washington University Press.

Index